Australian Signpost MATHS

1

Alan McSeveny
Diane McSeveny-Foster
Erika Johnson
Alan Parker
Rachel McSeveny

Pearson Australia
(a division of Pearson Australia Group Pty Ltd)
459–471 Church Street, Level 1, Building B, Richmond, Victoria 3121
PO Box 23360, Melbourne, Victoria 8012
www.pearson.com.au

First published 2018 by Pearson Australia
2025 2024 2023 2022
10 9 8 7 6 5 4 3 2 1

Content Learning Specialist: Julian Lumb
Project Manager: Shelly Wang
Production Manager: Elizabeth Gosman & Aptara
Editor: Aptara
Designer: Anne Donald & Jennifer Johnston
Cover design: Jennifer Johnston
Desktop operator: Aptara
Rights & Permissions Editor: Samantha Russell-Tulip
Senior Publishing Services Analyst: Rob Curulli
Cover art: Michael Barter
Cover background: rf_vector/shutterstock.com
Illustrators: Michael Barter, Wendy Gorton, Nives Porcellato, Andrew Craig
Printed in Australia by the SOS Print + Media Group

ISBN 978 1 4886 2178 9
Pearson Australia Group Pty Ltd ABN 40 004 245 943

Acknowledgements
All material identified by ACARA is material subject to copyright under the *Copyright Act 1968* and is owned by the Australian Curriculum, Assessment and Reporting Authority 2018. ACARA neither endorses nor verifies the accuracy of the information provided and accepts no responsibility for incomplete or inaccurate information. In particular, ACARA does not endorse or verify that:

- The content descriptions are solely for a particular year and subject;
- All the content descriptions for that year and subject have been used; and
- The author's material aligns with the Australian Curriculum content descriptions for the relevant year and subject.

You can find the unaltered and most up to date version of this material at http://www.australiancurriculum.edu.au/ This material is reproduced with the permission of ACARA.

Every effort has been made to trace and acknowledge copyright. However, if any infringement has occurred, the publishers tender their apologies and invite the copyright holders to contact them.

Some of the images used in *Australian Signpost Maths 1* might have associations with deceased Indigenous Australians. Please be aware that these images might cause sadness or distress in Aboriginal or Torres Strait Islander communities.

What is Australian Signpost Maths?

Australian Signpost Maths is a mathematics activity book series for students from Foundation to Year 6. The series has been written to meet the requirements of the Australian Curriculum.

The components of the series include Student Books, Teacher's Books, Mentals Books and an interactive Website. Teachers can select an appropriate program for every student from the rich and varied material provided.

The content has been carefully sequenced within each year level and across the series to take into account students' likely mathematical development.

Student Books

Teacher's Books

Mentals Books

Website

Structure of Australian Signpost Maths

Australian Signpost Maths emphasises the curriculum's syllabus content as well as problem-solving strategies, language development and the use of technology.

To maximise the benefits of the program, the Student Book, Teacher's Book, Mentals Book and Website should be used together.

The sequence of units in the **Student Book** forms a suggested program for the year. The **Teacher's Book** also provides lesson plans for each page of the Student Book, and blackline masters to assist teachers in implementing the program.

The Student Book presents lessons as a mix of content strands. However, the Contents and Contents Cross-reference pages in the Student Book allow teachers to construct programs based on the specific content strands (Number and Algebra, Measurement and Geometry, and Statistics and Probability). Progress Tests and remediation records are located in the Teacher's Book and on the website. These tests are also now included in the back of this book.

The **Mentals Book** mixes examples from all strands, reviewing the content of previous units of the Student Book.

The innovative **Website** help teachers to bring mathematics alive with technology. The website provides interactive maths tools, games and practice opportunities as well as relevant resource masters and worksheets for all year levels. These can be used for whole-class, small-group and individual learning. The website also includes **Concept Check-In**, a new diagnostic screener.

Special Features of Australian Signpost Maths

- **Traffic Light** system allows students to reflect on their work and highlight any units that they are having trouble understanding. They tick the red for units they feel they still don't understand, and green for those they feel they understand fully.
- Exercises are **well graded**. New work is reinforced in the Mentals Book.
- The **Progress Tests** (now also in the back of this book) allow the teacher to discover each student's strengths and weaknesses, and the cross-references direct students to the pages where that work is introduced.
- **Concept Check-In** diagnostic screener (on the Website) provides a snapshot of the class' conceptual understandings to aid in classroom management. It also allows teachers to measure progress over time..
- **Answers** are supplied in the Teacher's Book.
- The **Dictionary** at the beginning of this Student Book will help students to learn the language of mathematics.
- **ID Cards** (in the Mentals Book, Teacher's Book and Website) review the language of mathematics by asking students to identify common terms, shapes and symbols.
- Important **rules and concepts** are clearly highlighted.
- **Worked examples** and explanations are given throughout the Student Book where new ideas are introduced.
- The use of **colour** makes emphasis clear and is highly motivating.
- **Cartoons** give instruction and friendly advice.
- **Interactive Activities** are provided for whole-class, small-group and individual learning.

Australian Signpost Icons

Signpost icons are used throughout the book as cues to the essential nature of exercises and activities, and as a guide to ways of engaging with them. These icons often indicate alternative or more concrete approaches to dealing with concepts.

This icon highlights **important rules and concepts** occurring throughout the book. It often appears with worked examples.

Activities provide **applications and enrichment**. These activities usually involve the use of concrete materials and partner or group work.

These enjoyable activities are used to **motivate and involve** students in mathematical pursuits. They usually involve games and puzzles.

Investigations allow students to **explore and discover** maths concepts.

This icon indicates the use of computers, calculators or other **information and communications technology**.

Australian Curriculum Proficiency Strands

The proficiency strands of the Australian Curriculum describe how content is explored or developed – that is, the 'thinking and doing' of mathematics.

Understanding

Learning the concepts

*Students build a robust knowledge of adaptable and transferable mathematical concepts. They make connections between related concepts and progressively apply the familiar to develop new ideas. They develop an understanding of the relationship between the 'why' and the 'how' of mathematics.**

Conceptual understanding of maths ideas includes the explanation of a concept using text and diagrams. This occurs throughout Australian Signpost Maths at the top of many pages and is indicated by the Concept icon.

Fluency

Using the concepts

*Students develop skills in choosing appropriate procedures, carrying out procedures flexibly, accurately, efficiently and appropriately, and recalling factual knowledge and concepts readily.**

The practice of maths skills to build fluency occurs on every page of Australian Signpost Maths.

Problem Solving

Applying concepts and strategies to develop solutions to problems

*Students develop the ability to make choices, interpret, formulate, model and investigate problem situations, and communicate solutions effectively.**

Problem solving provides opportunities for students to use strategies and skills such as investigating and questioning, to collaborate with others and to communicate their findings to different audiences. Such activities are often indicated throughout Australian Signpost Maths by the Activity and Investigation icons.

Reasoning

Coherent and logical thought

*Students develop an increasingly sophisticated capacity for logical thought and actions, such as analysing, proving, evaluating, explaining, inferring, justifying and generalising.**

Students require opportunities to explain their mathematical thinking and can do so through both diagrams and written explanations. Reasoning questions are located throughout Australian Signpost Maths.

*The Australian Curriculum: Mathematics, v1.2 – Content structure

1 Contents and Syllabus Overview

KEY

Blue	Number & Algebra
Green	Measurement & Geometry
Orange	Statistics & Probability

Page	Unit	Title	Strand: Number and Algebra	Strand: Measurement and Geometry	Strand: Statistics and Probability	Sub-strand: Number and place value	Sub-strand: Fractions and decimals	Sub-strand: Money and financial mathematics	Sub-strand: Patterns and algebra	Sub-strand: Using units of measurement	Sub-strand: Shape	Sub-strand: Location and transformation	Sub-strand: Chance	Sub-strand: Data representation and interpretation	Suggested progress
1		Thinking Skills													Term 1
2	1A	Number Revision	■			●									
3	1B	Number Revision	■			●									
4	1C	Numbers 11 and 12	■			●									
5	1D	Numbers 11 to 16	■			●									
6	2A	Adding Two Groups	■			●									
7	2B	Addition Sentences	■			●									
8	2C	Numbers to 20	■			●									
9	2D	Numbers to 20	■			●									
10	3A	Ordinal Numbers	■			●									
11	3B	Subtraction	■			●									
12	3C	Analogue Time		■						●					
13	3D	Shapes Revision		■							●				
14	4A	Numbers to 20	■			●									
15	4B	Numbers to 20	■			●									
16	4C	Identifying Objects		■							●				
17	4D	3D Objects		■							●				
18	5A	Addition Sentences	■			●									
19	5B	Friends of 10	■			●									
20	5C	One Half	■				●								
21	5D	Introducing Graphs			■									●	
22	6A	Numbers to 20	■			●									T1*
23	6B	Subtraction	■			●									
24	6C	Patterns	■			●			●						
25	6D	Digital Time		■						●					
26	7A	Partitioning to 10	■			●									
27	7B	Sorting and Classifying Coins	■					●							
28	7C	Shapes in Our World		■							●				
29	7D	Using Pictures in Graphs			■									●	
30	8A	Addition Facts	■			●									Term 2
31	8B	Doubles	■			●									
32	8C	Units of Length		■						●					
33	8D	Informal Units of Length		■						●					
34	9A	Difference	■			●									
35	9B	Difference	■			●									
36	9C	Comparing Mass		■						●					
37	9D	The Oval and Other Shapes		■							●				

Page	Unit	Title	Strand: Number and Algebra	Strand: Measurement and Geometry	Strand: Statistics and Probability	Sub-strand: Number and place value	Sub-strand: Fractions and decimals	Sub-strand: Money and financial mathematics	Sub-strand: Patterns and algebra	Sub-strand: Using units of measurement	Sub-strand: Shape	Sub-strand: Location and transformation	Sub-strand: Chance	Sub-strand: Data representation and interpretation	Suggested progress
38	10A	Groups of 10	●			●									
39	10B	Place Value	●			●									
40	10C	Australian Coins	●					●							
41	10D	Left and Right		●								●			
42	11A	Subtraction	●			●									
43	11B	Difference Between Groups	●			●									
44	11C	Position Language		●								●			
45	11D	Location Language		●								●			
46	12A	Groups of 10	●			●									
47	12B	Counting by 10s	●			●									
48	12C	Difference	●			●									
49	12D	The Hexagon		●							●				
50	13A	Counting On	●			●									T2*
51	13B	Shape Hunt		●							●				
52	13C	Time Revision		●						●					
53	13D	Analogue and Digital Time		●						●					
54	14A	Numbers to 100	●			●									
55	14B	Numbers to 100	●			●									
56	14C	Counting Back	●			●									
57	14D	Drawing Shapes		●							●				
58	15A	The Number Line	●			●									
59	15B	The Calculator	●			●									
60	15C	Informal Units of Capacity		●						●					
61	15D	Informal Units of Capacity		●						●					
62	16A	Numbers to 20	●			●									
63	16B	Addition to 20	●			●									
64	16C	Addition to 20	●			●									
65	16D	Chance Words			●								●		
66	17A	Numbers to 100	●			●									Term 3
67	17B	Numbers to 100	●			●									
68	17C	Addition to 20	●			●									
69	17D	Recognising 3D Objects		●							●				
70	18A	Numbers to 100	●			●									
71	18B	Partitioning	●			●									
72	18C	Partitioning	●			●									
73	18D	Half Past		●						●					
74	19A	Subtraction to 20	●			●									
75	19B	Subtraction by Counting On	●			●									
76	19C	One Half	●				●								
77	19D	Half of a Group	●				●								
78	20A	Numbers to 100	●			●									
79	20B	Counting by 2s, 5s and 10s	●			●			●						
80	20C	Looking at Patterns	●						●						
81	20D	Half Past		●						●					
82	21A	Number Relationships	●			●									
83	21B	Adding Equal Groups	●			●									

Page	Unit	Title	Strand	Number and Algebra	Measurement and Geometry	Statistics and Probability	Sub-strand	Number and place value	Fractions and decimals	Money and financial mathematics	Patterns and algebra	Using units of measurement	Shape	Location and transformation	Chance	Data representation and interpretation	Suggested progress
84	21C	Graphing Data Using Blocks				■										●	
85	21D	Graphs				■										●	
86	22A	Partitioning		■				●									T3*
87	22B	Partitioning		■				●									
88	22C	Informal Units of Length			■							●					
89	22D	Informal Units of Length			■							●					
90	23A	Place Value		■				●									
91	23B	Place Value		■				●									
92	23C	The Rhombus			■								●				
93	23D	2D Shapes			■								●				
94	24A	Numbers to 100		■				●									
95	24B	Place Value		■				●									
96	24C	Patterns		■				●			●						
97	24D	Months of the Year			■							●					
98	25A	Counting On		■				●									
99	25B	Addition by Counting On		■				●									
100	25C	Digital and Analogue Time			■							●					
101	25D	Making a Graph				■										●	
102	26A	Counting Back		■				●									Term 4
103	26B	Subtraction		■				●									
104	26C	Properties of Shapes			■								●				
105	26D	The Cube			■								●				
106	27A	Subtraction by Counting On		■				●									
107	27B	Addition		■				●									
108	27C	One Half		■					●								
109	27D	Giving Directions			■									●			
110	28A	Place Value		■				●									
111	28B	Relating Addition and Subtraction		■				●									
112	28C	Comparing Capacities			■							●					
113	28D	Comparing Capacities			■							●					
114	29A	Addition		■				●									T4*
115	29B	How Many More?		■				●									
116	29C	Giving Directions			■									●			
117	29D	Chance Language				■									●		
118	30A	Inverse Operations		■				●									
119	30B	Number Patterns		■				●			●						
120	30C	Duration Using Weeks			■							●					
121	30D	Duration Using Hours			■							●					
122	31A	The Calculator		■				●									
123	31B	Calculators		■				●			●						
124	31C	Mass			■							●					
125	31D	Dot Patterns			■								●				
126	32A	Ordinal Numbers		■				●									
127		Appendix: Australian Money															

*Suggested placement for Progress Tests 1 to 4 (see the Teacher's Book). It is assumed that there are 10 weeks in each term.

Number and Algebra

1 Counting	Pages	Australian Curriculum Reference
Count to and from any starting point up to 100	2, 3, 4, 5, 8, 9, 22, 54, 58, 62, 66, 94	Develop confidence with number sequences to and from 100 by ones from any starting point, and skip count by twos, fives and tens starting from zero (ACMNA012); Recognise, model, read, write and order numbers to at least 100, and locate these numbers on a number line (ACMNA013); Count collections to 100 by partitioning numbers using place value (ACMNA014)
Understand and reason with number sequences to and from 100	9, 22, 47, 58, 62, 66, 67, 79, 94	Develop confidence with number sequences to and from 100 by ones from any starting point, and skip count by twos, fives and tens starting from zero (ACMNA012); Recognise, model, read, write and order numbers to at least 100, and locate these numbers on a number line (ACMNA013); Count collections to 100 by partitioning numbers using place value (ACMNA014)
Say number sequences of twos, fives and tens starting from zero	47, 67, 70, 79, 80, 94	Develop confidence with number sequences to and from 100 by ones from any starting point, and skip count by twos, fives and tens starting from zero (ACMNA012)
Use a calculator to increase understanding of counting patterns	54, 82, 119, 122, 123	Develop confidence with number sequences to and from 100 by ones from any starting point, and skip count by twos, fives and tens starting from zero (ACMNA012)
Ordinal numbers	10, 126	Develop confidence with number sequences to and from 100 by ones from any starting point, and skip count by twos, fives and tens starting from zero (ACMNA012)
2 Numeration		
Model and represent numbers and the use of place-value cards	2, 3, 4, 8, 9, 14, 15, 38, 46, 54, 55, 90, 91, 94, 95	Recognise, model, read, write and order numbers to at least 100, and locate these numbers on a number line (ACMNA013); Count collections to 100 by partitioning numbers using place value (ACMNA014)
Read, write and order numbers to 100	2, 3, 4, 5, 8, 14, 15, 62, 66, 67, 70, 78, 90, 91, 94, 95	Develop confidence with number sequences to and from 100 by ones from any starting point, and skip count by twos, fives and tens starting from zero (ACMNA012); Recognise, model, read, write and order numbers to at least 100, and locate these numbers on a number line (ACMNA013); Count collections to 100 by partitioning numbers using place value (ACMNA014)
Use Base 10 materials	14, 38, 46, 55, 95	Develop confidence with number sequences to and from 100 by ones from any starting point, and skip count by twos, fives and tens starting from zero (ACMNA012); Recognise, model, read, write and order numbers to at least 100, and locate these numbers on a number line (ACMNA013); Count collections to 100 by partitioning numbers using place value (ACMNA014)
Read numerals on a calculator	46, 54, 82, 110, 119	Develop confidence with number sequences to and from 100 by ones from any starting point, and skip count by twos, fives and tens starting from zero (ACMNA012); Recognise, model, read, write and order numbers to at least 100, and locate these numbers on a number line (ACMNA013)
Order numbers in sequences	22, 62, 66, 91	Develop confidence with number sequences to and from 100 by ones from any starting point, and skip count by twos, fives and tens starting from zero (ACMNA012); Recognise, model, read, write and order numbers to at least 100, and locate these numbers on a number line (ACMNA013)

3 Place value		
Count and record numbers by grouping in tens	9, 22, 39, 54, 55, 90, 91, 94, 95, 110	Develop confidence with number sequences to and from 100 by ones from any starting point, and skip count by twos, fives and tens starting from zero (ACMNA012); Recognise, model, read, write and order numbers to at least 100, and locate these numbers on a number line (ACMNA013); Count collections to 100 by partitioning numbers using place value (ACMNA014)
Partition and regroup numbers	9, 26, 71, 72, 82, 86, 87, 90, 95, 110	Count collections to 100 by partitioning numbers using place value (ACMNA014); Represent and solve simple addition and subtraction problems using a range of strategies including counting on, partitioning and rearranging parts (ACMNA015)
4 Fractions		
One half as one of two equal parts	20, 76	Recognise and describe one-half as one of two equal parts of a whole (ACMNA016)
Halves of collections	76, 77, 108	Recognise and describe one-half as one of two equal parts of a whole (ACMNA016)
5 Addition and subtraction		
Model, represent and solve problems concerning addition	4, 6, 7, 18, 19, 26, 30, 31, 50, 59, 63, 64, 68, 71, 72, 82, 83, 86, 87, 96, 98, 99, 107, 111, 114, 115, 118, 122	Represent and solve simple addition and subtraction problems using a range of strategies including counting on, partitioning and rearranging parts (ACMNA015)
Model, represent and solve problems concerning subtraction	11, 23, 34, 35, 42, 48, 56, 74, 75, 96, 102, 103, 111, 115, 118	Represent and solve simple addition and subtraction problems using a range of strategies including counting on, partitioning and rearranging parts (ACMNA015)
Model, represent and solve problems concerning grouping	83, 123	Represent and solve simple addition and subtraction problems using a range of strategies including counting on, partitioning and rearranging parts (ACMNA015)
Connect addition and subtraction	75, 106, 111, 115, 118	Represent and solve simple addition and subtraction problems using a range of strategies including counting on, partitioning and rearranging parts (ACMNA015)
Problems involving a missing element	72, 75, 106, 107, 114, 122	Represent and solve simple addition and subtraction problems using a range of strategies including counting on, partitioning and rearranging parts (ACMNA015)
Subtraction (taking away)	11, 42, 74	Represent and solve simple addition and subtraction problems using a range of strategies including counting on, partitioning and rearranging parts (ACMNA015)
Subtraction (difference)	34, 35, 43, 48	Represent and solve simple addition and subtraction problems using a range of strategies including counting on, partitioning and rearranging parts (ACMNA015)
Mental strategies (count on, doubles, make to ten)	31, 50, 56, 59, 71, 72, 74, 75, 82, 86, 87, 98, 99, 102, 103, 106, 118	Represent and solve simple addition and subtraction problems using a range of strategies including counting on, partitioning and rearranging parts (ACMNA015)
6 Number patterns		
Patterns with objects	24, 54, 80	Investigate and describe number patterns formed by skip counting and patterns with objects (ACMNA018)
Patterns with numbers	4, 24, 47, 70, 79, 80, 96, 119, 123	Develop confidence with number sequences to and from 100 by ones from any starting point, and skip count by twos, fives and tens starting from zero (ACMNA012); Investigate and describe number patterns formed by skip counting and patterns with objects (ACMNA018)
7 Money		
Recognise, describe and order Australian coins	27, 40, 101	Recognise, describe and order Australian coins according to their value (ACMNA017)

Measurement and Geometry

1 Length, capacity and mass	Pages	Australian Curriculum Reference
Use uniform informal units to measure length	32, 33, 88, 89	Measure and compare the lengths and capacities of pairs of objects using uniform informal units (ACMMG019)
Use uniform informal units to measure capacity	60, 61, 112, 113	Measure and compare the lengths and capacities of pairs of objects using uniform informal units (ACMMG019)
Compare lengths	32, 89	Measure and compare the lengths and capacities of pairs of objects using uniform informal units (ACMMG019)
Compare capacity	61, 112, 113	Measure and compare the lengths and capacities of pairs of objects using uniform informal units (ACMMG019)
Comparing mass	36, 124	Use direct and indirect comparisons to decide which is longer, heavier or holds more, and explain reasoning in everyday language (ACMMG006) *[Progression]*
2 Time		
Read time to the hour and half hour on digital and analogue clocks	12, 25, 52, 53, 73, 81, 100	Tell time to the half-hour (ACMMG020)
Make connections between common time sequences such as days of the week, months	97, 120	Describe duration using months, weeks, days and hours (ACMMG021)
Duration of time: months, weeks, days, hours	97, 120, 121	Describe duration using months, weeks, days and hours (ACMMG021)
3 Geometry		
Recognise, visualise and classify familiar 2D shapes	13, 37, 49, 57, 69, 92, 93, 104, 125	Recognise and classify familiar two-dimensional shapes and three-dimensional objects using obvious features (ACMMG022)
Draw 2D shapes	13, 28, 57, 92, 93, 125	Recognise and classify familiar two-dimensional shapes and three-dimensional objects using obvious features (ACMMG022)
Features of 2D shapes	37, 49, 57, 92, 93, 104	Recognise and classify familiar two-dimensional shapes and three-dimensional objects using obvious features (ACMMG022)
Recognise, visualise and classify familiar 3D objects	16, 17, 28, 51, 69, 105	Recognise and classify familiar two-dimensional shapes and three-dimensional objects using obvious features (ACMMG022)
Make models of 3D objects	51	Recognise and classify familiar two-dimensional shapes and three-dimensional objects using obvious features (ACMMG022)
Features of 3D objects	69, 105	Recognise and classify familiar two-dimensional shapes and three-dimensional objects using obvious features (ACMMG022)
4 Location		
Give and follow directions	41, 45, 109, 116	Give and follow directions to familiar locations (ACMMG023)
Language of location	41, 44, 45, 109, 116	Give and follow directions to familiar locations (ACMMG023)

Statistics and Probability

1 Data representation	Pages	Australian Curriculum Reference
Collecting data (class data)	84, 85, 88	Choose simple questions and gather responses (ACMSP262)
Representing data	84, 85	Represent data with objects and drawings where one object or drawing represents one data value, and describe the displays (ACMSP263)
Using and understanding picture graphs and other graphs	21, 29, 84, 85	Represent data with objects and drawings where one object or drawing represents one data value, and describe the displays (ACMSP263)
2 Data interpretation		
Read and make connections between lists, tables and picture graphs	21, 29, 85, 88, 101	Choose simple questions and gather responses (ACMSP262); Represent data with objects and drawings where one object or drawing represents one data value, and describe the displays (ACMSP263)
Convey the story told on a graph (draw conclusions) and make statements about data	21, 29, 84, 85, 101	Choose simple questions and gather responses (ACMSP262); Represent data with objects and drawings where one object or drawing represents one data value, and describe the displays (ACMSP263)
3 Chance		
Possible outcomes from a chance event	117	Identify outcomes of familiar events involving chance and describe them using everyday language such as 'will happen', 'won't happen' or 'might happen' (ACMSP024)
Likely and unlikely events	65, 117	Identify outcomes of familiar events involving chance and describe them using everyday language such as 'will happen', 'won't happen' or 'might happen' (ACMSP024)

capacity

Capacity is the amount that a container can hold.

- A container can be **empty**.
- A container can be **full**.
- A container can be **half full**.

coins

5c 10c 20c 50c $1 $2

collection

A collection is a group of objects.

column

A column is a line of objects going down.
Here are 5 columns of 2.

curved line

A curved line is not straight.

double

Double means the same thing twice.

Double 4 = 4 + 4 = 8

equals sign =

The equals sign means 'makes' or 'is equal to'.

graph

A graph shows information (or 'data').
This graph is a picture graph.

half

One half is one of two equal parts.

This orange is cut in half.

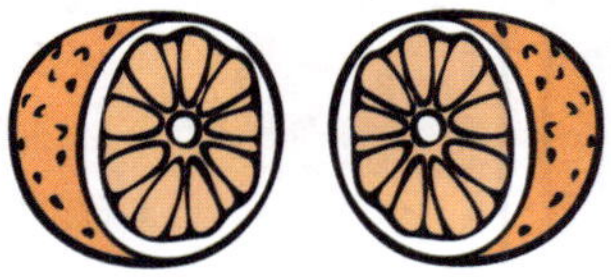

left and right

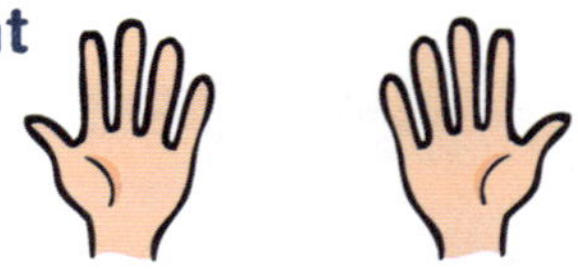

- Left
- Right

length words:

deeper	long	tall
distance	longer	taller
higher	short	thicker
lower	shorter	thinner

list

A list is a set of information. Lists can be used to show data that is collected.

mass words:

heavy	heaviest	lighter
heavier	light	lightest

number line

A number line is a line that shows numbers in order. Number lines can be used for counting and adding.

numeral

A numeral is a written number symbol such as 0, 7, 13 or 100.

ordinal number

Ordinal numbers describe the order or position of something, for example 1st, 2nd, 3rd, 4th, 15th or 31st.

partitioning

Partitioning is when a group of objects is broken up into two parts. The more objects there are in the group, the more different combinations of groups can be made.
For example, the number 10 can be partitioned into 1 + 9, 2 + 8, 3 + 7, 4 + 6 and so on.

pattern

A pattern is a group of numbers, objects, shapes, colours, sounds or actions that are repeated over and over again.

picture graph

See *graph*.

place value

row

A row is a line of objects going across. Here are 2 rows of 5.

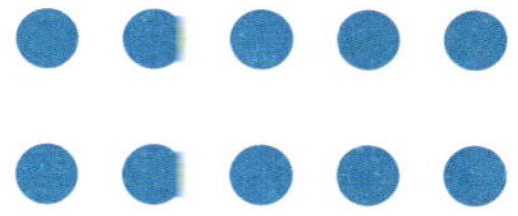

right and left

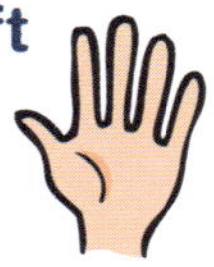

- Left

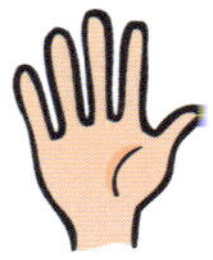

- Right

sharing

When sharing, we make sure that each share is the same size.

2 people could share these 6 balls. Each person would get 3 balls.

straight line

A straight line has no bends or curves.

table

A table is a set of information in rows and columns. We use tables to organise information or data collected.

time words:

morning	day	afternoon	night

Days						
Sunday	Monday	Tuesday	Wednesday	Thursday	Friday	Saturday

Months					
January	February	March	April	May	June
July	August	September	October	November	December

- Clocks

3 o'clock

Half past 3

o'clock When the long hand is pointing to 12, the time is an o'clock time.

half past When the long hand is pointing to the 6 and the short hand is pointing in between two numbers, the time is a 'half past' time.

two-dimensional (2D) shapes

Flat shapes are two-dimensional.
They have length and width.

Circle
1 curved side

Triangle
3 sides

Square
4 equal sides

Rectangle
2 equal long sides and 2 equal short sides, like a stretched square

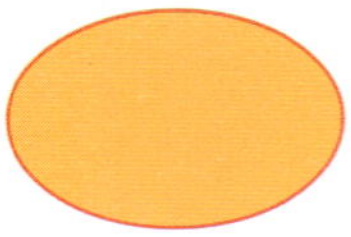

Oval
1 curved side,
like a squashed circle

Hexagon
6 sides and 6 corners

three-dimensional (3D) objects

Solid objects are three-dimensional.
They have length, width and height.

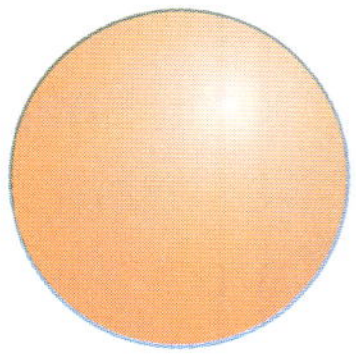

Ball-shaped objects (spheres)
are curved and round.
They can roll.

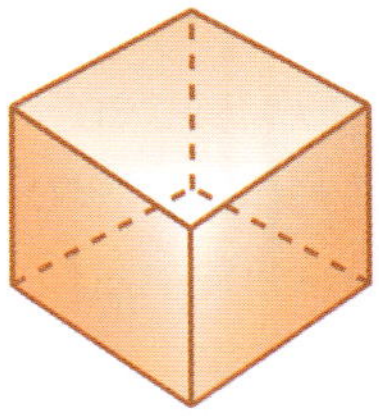

Box-shaped objects (cubes)
have 6 square faces.
They can slide.

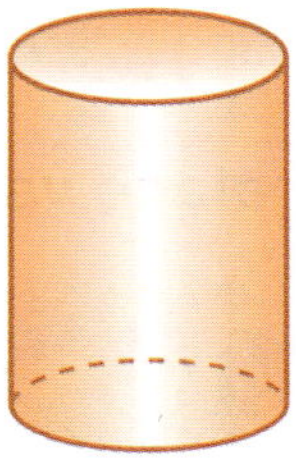

Can-shaped objects (cylinders)
have 2 circle faces and 1 curved surface. They can roll and slide.

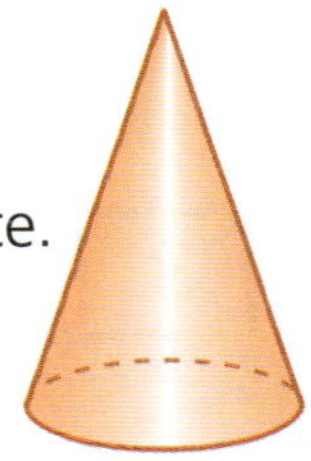

Cone-shaped objects (cones)
have 1 circle face and 1 curved surface.
They can roll and slide.

Numbers

0 zero
1 one
2 two
3 three
4 four
5 five
6 six
7 seven
8 eight
9 nine
10 ten

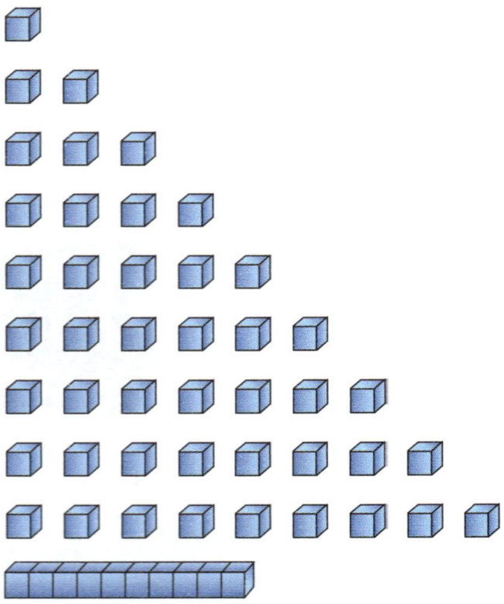

11 eleven
12 twelve
13 thirteen
14 fourteen
15 fifteen
16 sixteen
17 seventeen
18 eighteen
19 nineteen
20 twenty

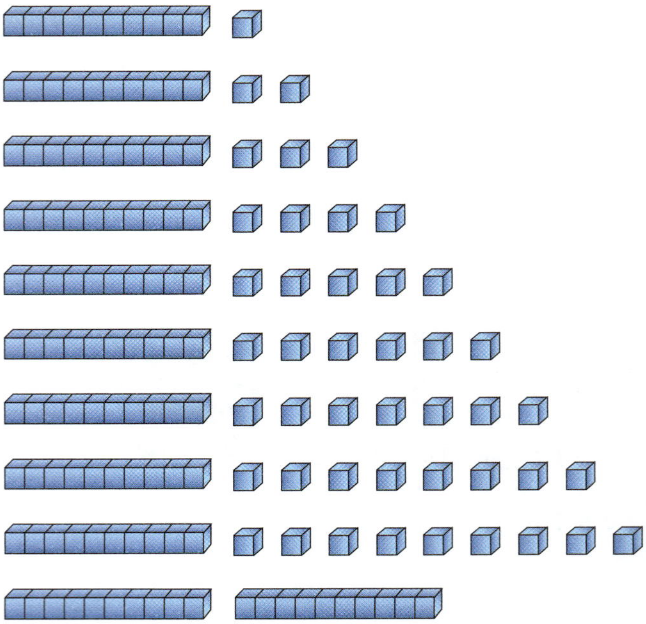

1 Thinking Skills

1. What can you count in this picture?
2. How many noses are in this picture?
3. Colour the round things red.
4. What are the mouse and the platypus doing?
5. The mouse is going to make a hat for its costume.
   Why should the mouse's hat have no ears?
6. What could the animals make with the paper?
7. How are the mouse and the platypus different?
8. Which of these questions have you liked best? Why?
9. Make up a maths question about this picture.

1A Number Revision

REVISION

CONCEPT

1	2	3	4	5
one	two	three	four	five

❶ Write the numeral and its name. Draw the number of balls.

1			one	
2				
3				
4				
5				

FUN SPOT

❷ Match each word with a numeral.

one	two	three	four	five
2	1	5	3	4

 • *AUSTRALIAN SIGNPOST MATHS 1* • ISBN 978 1 4886 2178 9

1B Number Revision

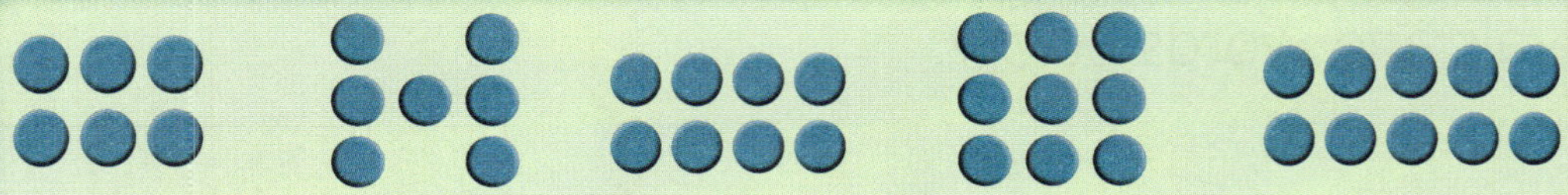

6	7	8	9	10
six	seven	eight	nine	ten

CONCEPT

6, 8 and 10 are even numbers.

1 Write the numeral and its name. Draw the number of hats.

6			six	
7				
8				
9				
10				

FUN SPOT

2 Match each word with a numeral.

six seven eight nine ten

8 6 7 10 9

 • *AUSTRALIAN SIGNPOST MATHS 1* • ISBN 978 1 4886 2178 9

1C Numbers 11 and 12

1 Trace the numerals and words below.

11 eleven   12 twelve

2 Count and write the number of objects.

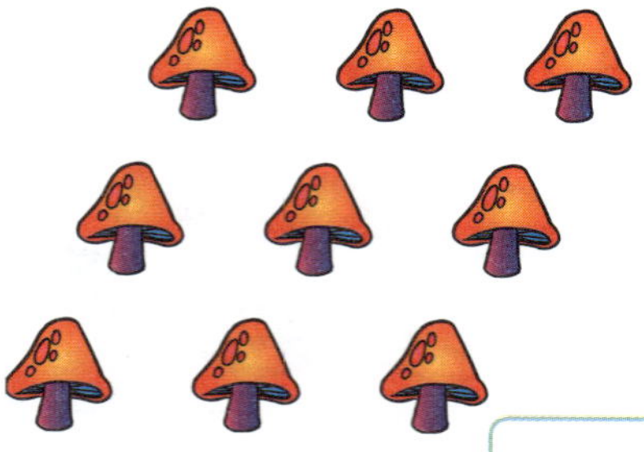

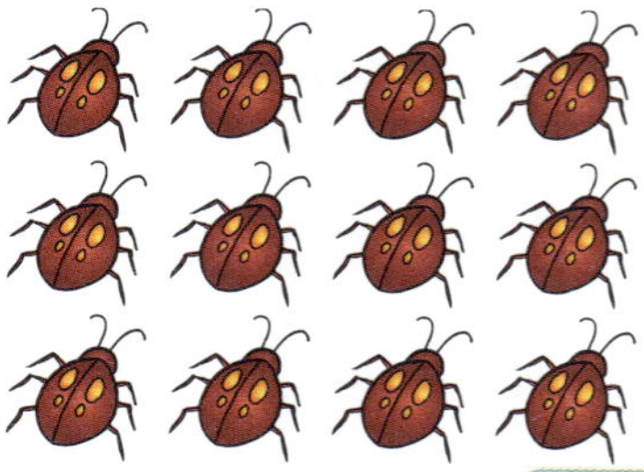

3 Count forwards and backwards. Colour every second number. Discuss.

1	2	3	4	5	6	7	8	9	10
11	12	13	14	15	16	17	18	19	20

4 Write the missing numbers.

9	10	
6	7	
7		9
10		12

5 Draw dots to make 11.
How many dots did you draw?

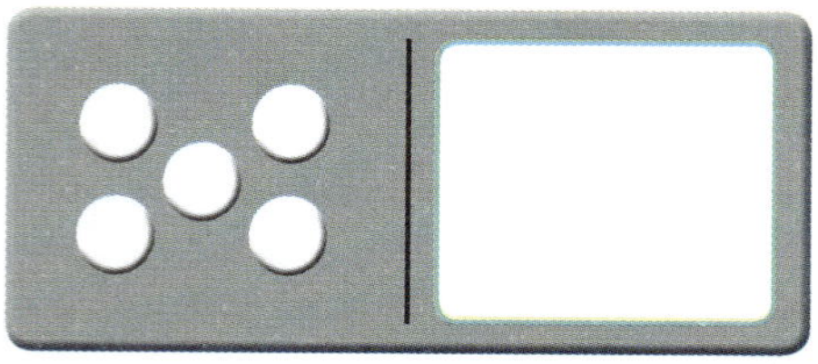

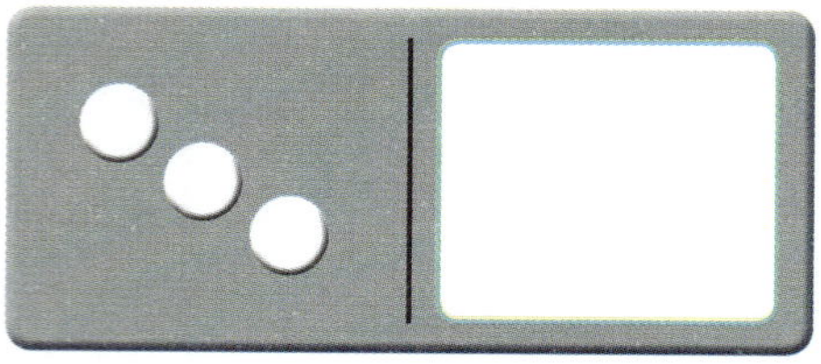

 • *AUSTRALIAN SIGNPOST MATHS 1* • ISBN 978 1 4886 2178 9

1D Numbers 11 to 16

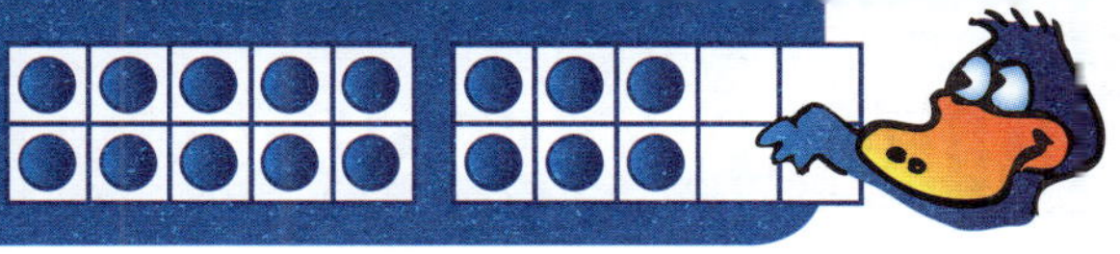

1 Trace over the numerals.

11 12 13 14 15 16

2 Count and write the number of objects.

3 Write the missing numbers.

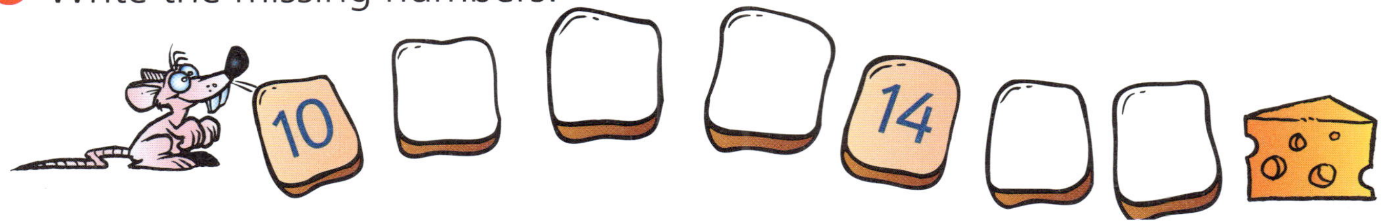

4 Trace the numerals and write the numbers.

13	13	
14	14	
16	16	

5 Colour the counting path from 1 to 16.

10	11	12	13	14	15	16	16
9	6	7	8	9	10	13	15
8	5	13	14	15	15	9	14
7	10	11	12	5	11	12	13
4	4	5	6	2	10	7	4
2	3	1	7	8	9	6	3
1	3	4	5	8	7	5	2

2A Adding Two Groups

1 a  4 and ☐ makes ☐.

b  ☐ and ☐ makes ☐.

c 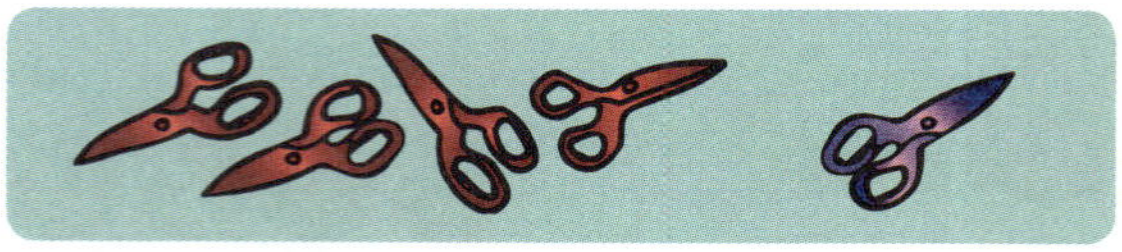 ☐ and ☐ makes ☐.

d  ☐ and ☐ makes ☐.

e  ☐ and ☐ makes ☐.

ACTIVITY

2 Draw your own pictures to complete the problems.

2 and 5 makes ☐.

7 and 3 makes ☐.

 • *AUSTRALIAN SIGNPOST MATHS 1* • ISBN 978 1 4886 2178 9

2B Addition Sentences

CONCEPT

+ means **add**, **and** or **plus**.

= means **makes** or **is equal to**.

5 + 3 = 8

1.  and  makes ☐ stars.

   ☐ + ☐ = ☐

2.  and  makes ☐ cakes.

   ☐ + ☐ = ☐

3.  and  makes ☐ balloons.

   ☐ + ☐ = ☐

4. 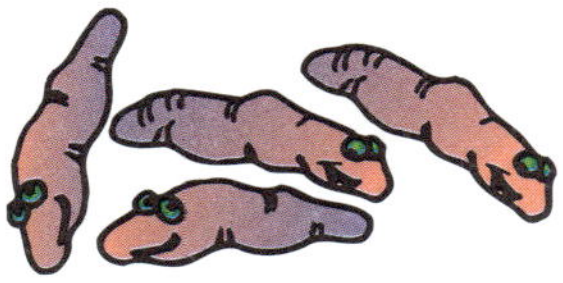 and 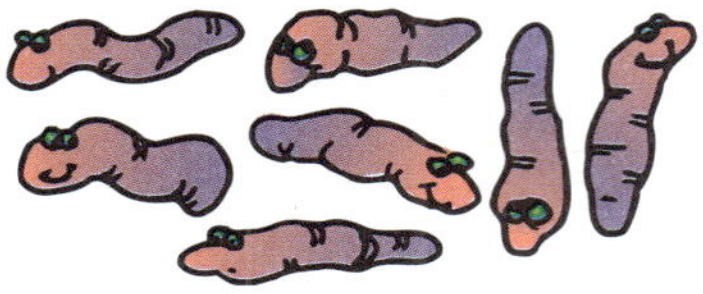 makes ☐ grubs.

   ☐ + ☐ = ☐

ACTIVITY

Use pictures to show number stories.

5 + 2 + 3 → 

20C Numbers to 20

20

1 Write the missing numbers. Count forwards and backwards.

1			4				8		
		13		15		17			

2 Trace each numeral and draw lines to match.

16

14

20

15

18

19

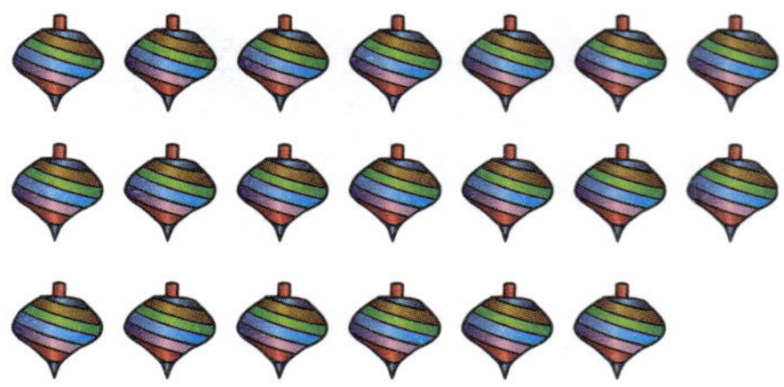

3 Draw 20 balls.

FUN SPOT

4 Write these numbers in order from largest to smallest:

16, 17, 18, 19, 20

→ How many students are in your class today?

 • *AUSTRALIAN SIGNPOST MATHS 1* • ISBN 978 1 4886 2178 9

2D Numbers to 20

CONCEPT

Numbers can be put into rows of 10. Count on to find each number.

1 Count the objects and complete.

10 and 3 is ☐.

10 and 6 is ☐.

10 and 9 is ☐.

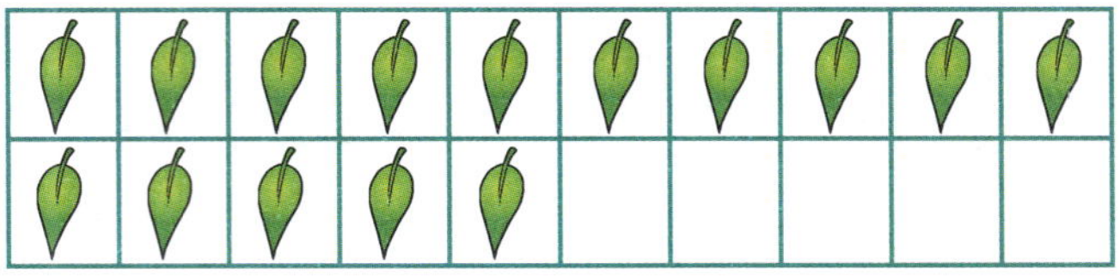

10 and 5 is ☐.

2 Write the missing numbers. Count by 5s and colour these numbers blue. Count by 10s and circle these numbers.

1			4	
		8		
11				
	17			

INVESTIGATION

3 Circle groups of 10.

☐ tens = ☐

3A Ordinal Numbers

1 Trace the ordinal numbers.

1st 2nd 3rd 4th 5th 6th

2 Write the position of each child. Who is last?

3 Write the ordinal numbers.

first	1st	sixth		eleventh	
second		seventh		twelfth	
third		eighth		thirteenth	
fourth		ninth		fourteenth	
fifth		tenth		fifteenth	

 • *AUSTRALIAN SIGNPOST MATHS 1* • ISBN 978 1 4886 2178 9

3B Subtraction

Take away is the same as **subtraction** or **minus**.

Cross out 2 dogs and complete the story.

8 take away 2 equals ☐.

1.  4 take away 1 equals ☐.
2. 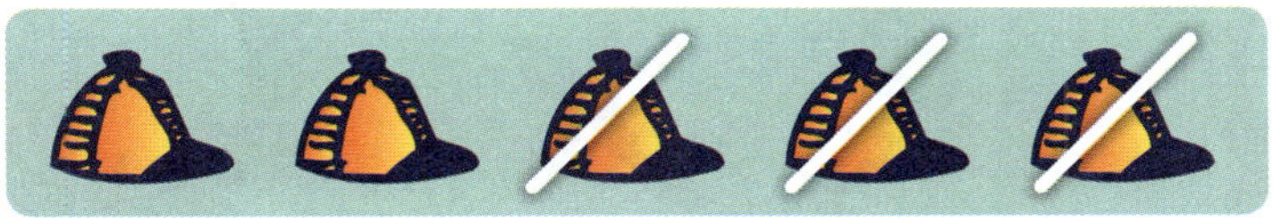 5 take away 3 equals ☐.
3.  6 take away 4 equals ☐.
4.  5 take away 5 equals ☐.
5. 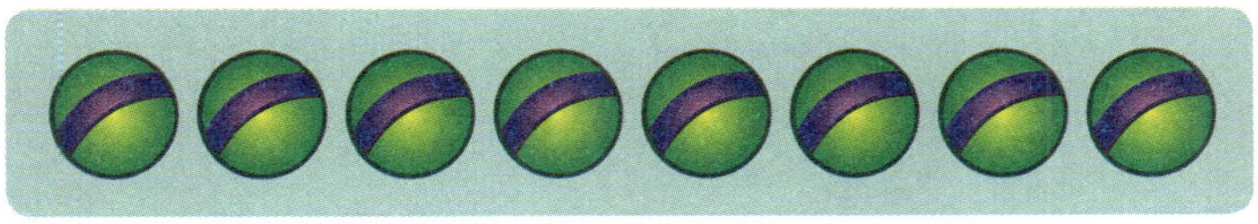 8 take away 6 equals ☐.
6.  9 take away 3 equals ☐.
7.  10 take away 5 equals ☐.

Work with a partner.

Use objects to model your own subtraction stories.

 • *AUSTRALIAN SIGNPOST MATHS 1* • ISBN 978 1 4886 2178 9

3C Analogue Time

1 Show the time.

a

7 o'clock

b

5 o'clock

c

2 o'clock

d

9 o'clock

e

10 o'clock

f

4 o'clock

g

12 o'clock

h

8 o'clock

2 Match each picture with the best time.

8 o'clock

10 o'clock

7 o'clock

4 o'clock

9 o'clock

12 o'clock

5 o'clock

6 o'clock

3D Shapes Revision

1 Colour the shapes in the picture:

red

yellow

green

blue

How many circles? ☐

How many squares? ☐

How many triangles? ☐

How many rectangles? ☐

2 Draw your own shapes picture.

© PEARSON AUSTRALIA 2018 • *AUSTRALIAN SIGNPOST MATHS 1* • ISBN 978 1 4886 2178 9

4A Numbers to 20

CONCEPT

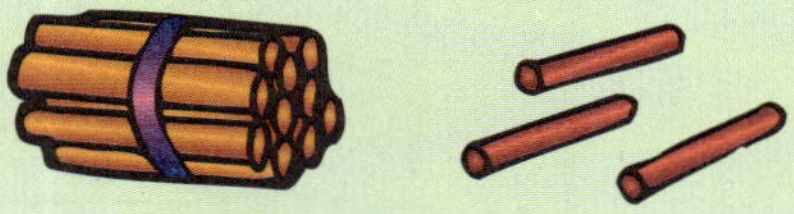

One ten and three ones.

1 Write the numbers shown.

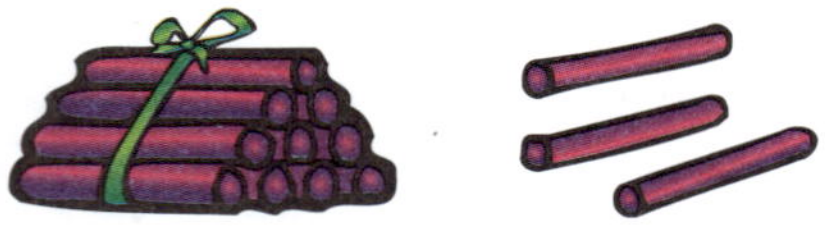 ☐

 ☐

 ☐

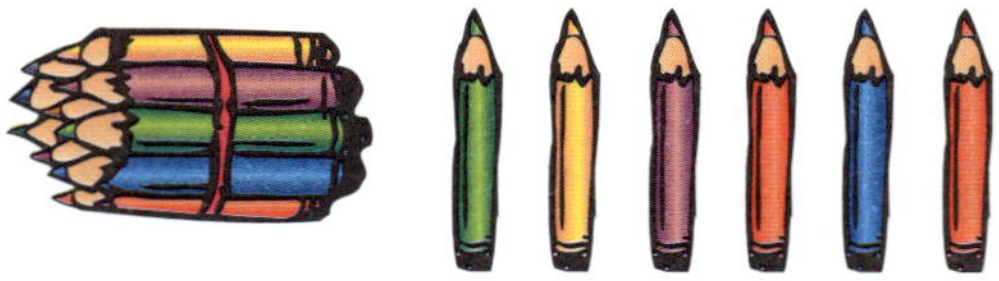 ☐

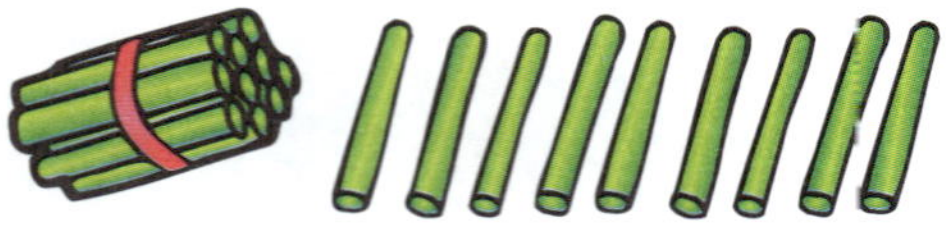 ☐

 ☐

2 Trace and write the numerals.

17 ☐, 18 ☐, 19 ☐, 20 ☐

3 Colour the blocks to match each number. Trace the numerals.

17 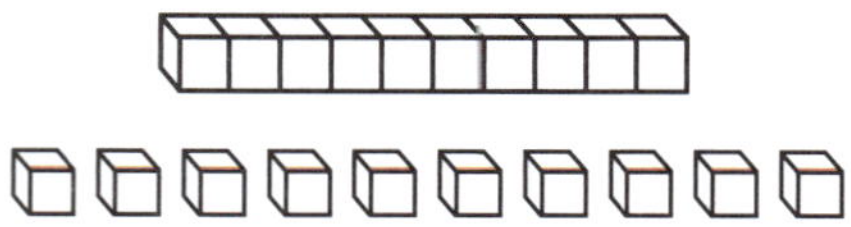

18 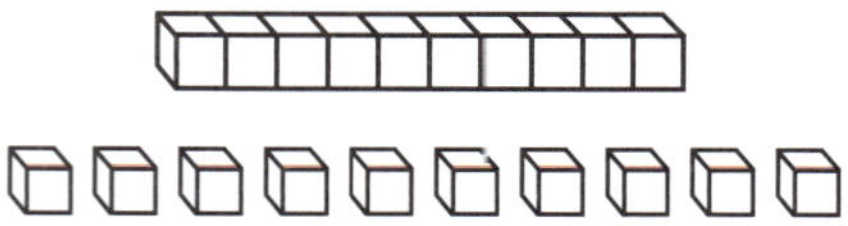

19 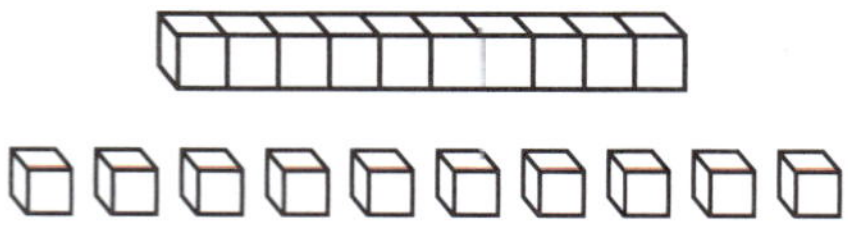

ACTIVITY

With a partner, take turns to make these numbers.

11, 18, 13, 20, 15

Use 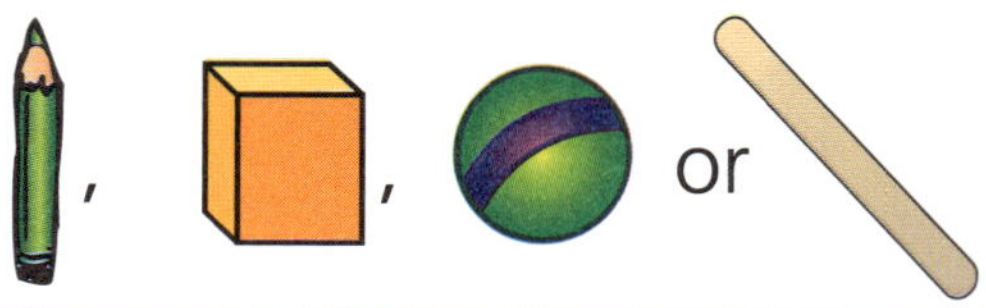 or

16 = 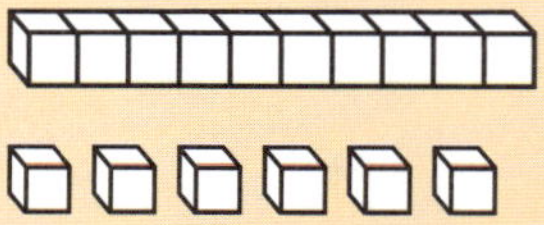

4B Numbers to 20

REVISION

CONCEPT

These are the numbers from 10 to 20.

1 Write the numbers shown.

a  ☐

b 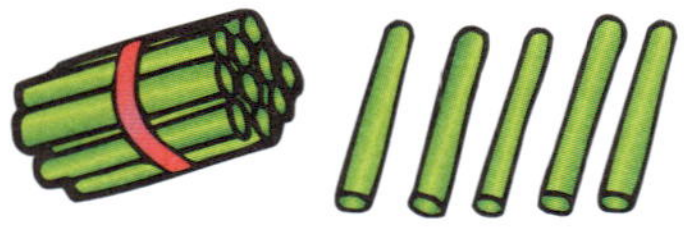 ☐

c 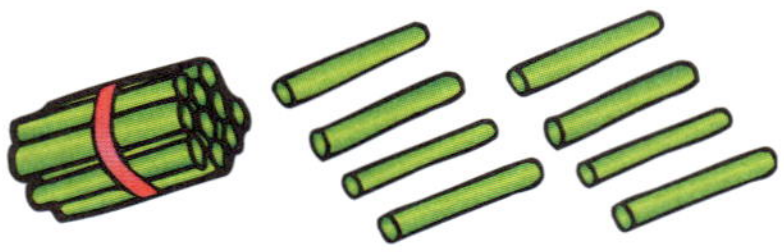 ☐

d 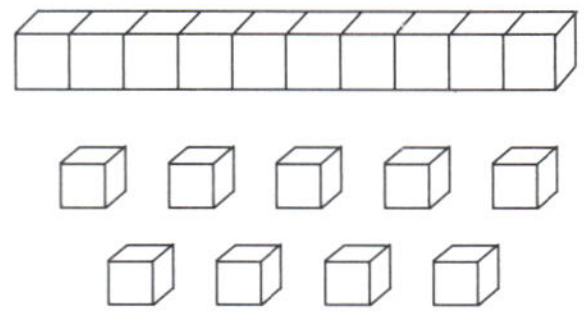 ☐

e 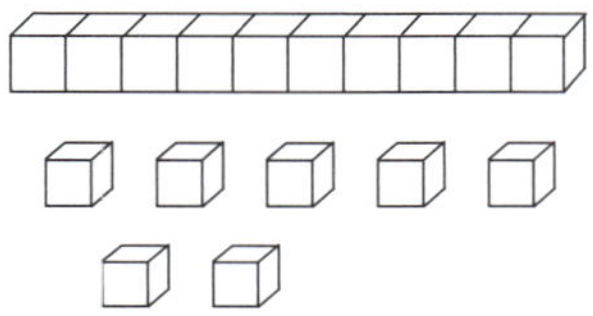 ☐

f  ☐

2 Write the numbers before and after.

a	☐	12	☐	**b**	☐	16	☐
c	☐	14	☐	**d**	☐	19	☐

INVESTIGATION

The Wurundjeri people used these symbols to stand for numbers.

Biceps

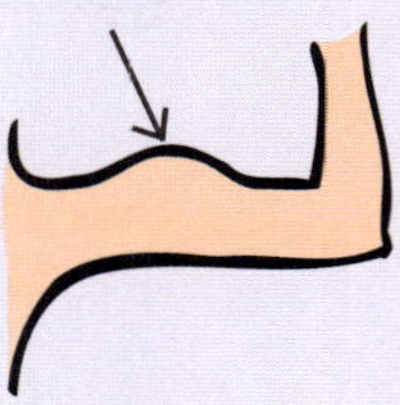

10

Neck

13

Earlobe

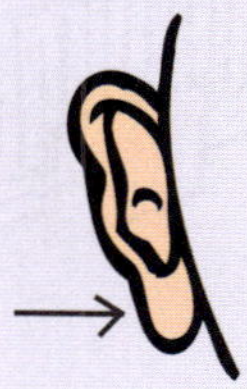

14

- Investigate other symbols that stand for numbers.

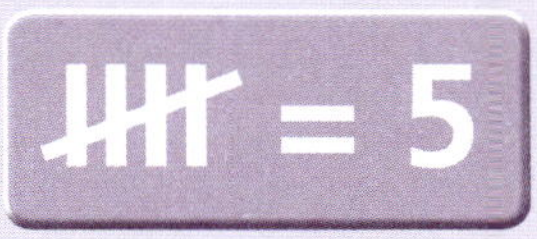

4C Identifying Objects

1 Colour in the picture, then count all the objects like these.

Can	Ball	Box	Curvy	Cone
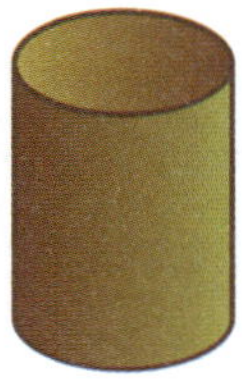				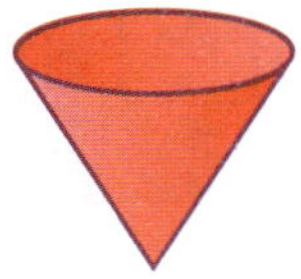

 • *AUSTRALIAN SIGNPOST MATHS 1* • ISBN 978 1 4886 2178 9

4D 3D Objects

1 Colour the objects that match the shape on the left.

 • *AUSTRALIAN SIGNPOST MATHS 1* • ISBN 978 1 4886 2178 9

5A Addition Sentences

1 Complete the number sentences.

a 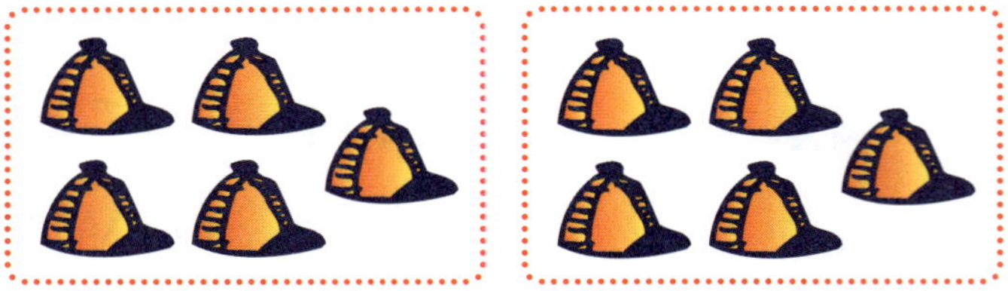 ☐ + ☐ = ☐

b  ☐ + ☐ = ☐

c  ☐ + ☐ = ☐

d  ☐ + ☐ = ☐

2 Make your own number sentence.

☐ + ☐ = ☐

5B Friends of 10

CONCEPT

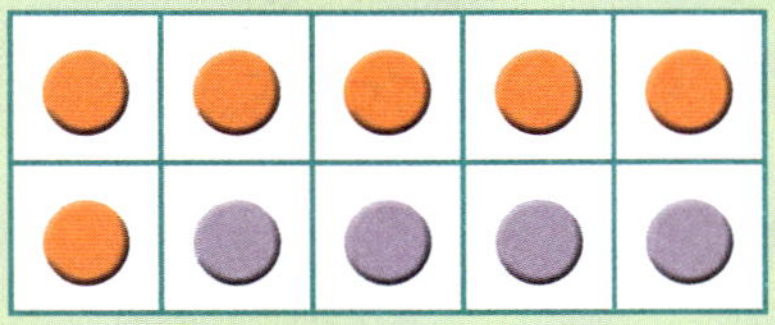

6 + 4 = 10

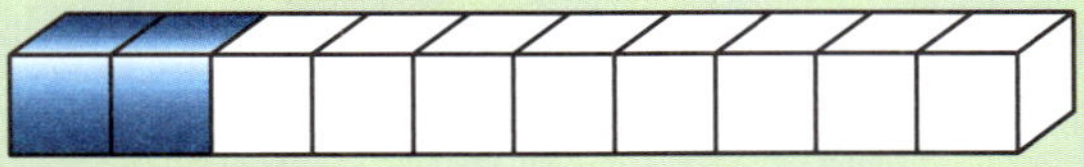

2 + 8 = 10

Learn the pairs that make 10.

1 Add dots so that each has a total of 10.

a
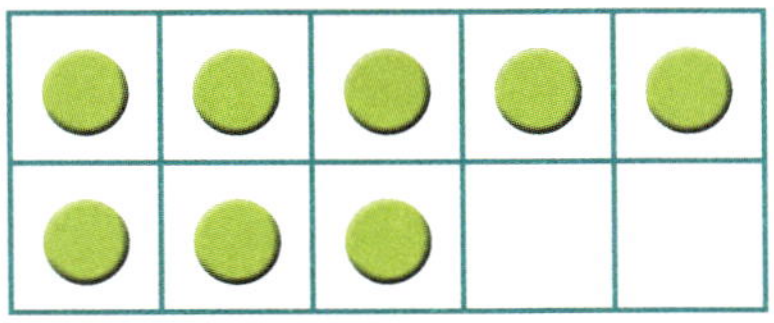

☐ + ☐

b
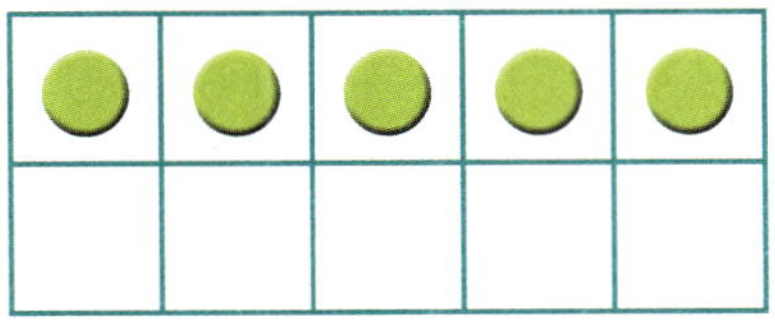

☐ + ☐

c
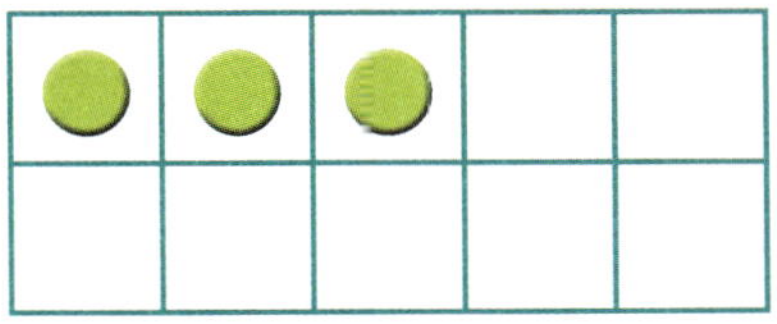

☐ + ☐

2 Use 10 counters to find all the possible ways we can add to make 10.

0 + 10

1 + ☐

2 + ☐

☐ + ☐

☐ + ☐

☐ + ☐

☐ + ☐

☐ + ☐

☐ + ☐

☐ + ☐

☐ + ☐

ACTIVITY

Match number cards 0 to 10 to make pairs of numbers that make 10.

- How many matches did you make? ☐

 • *AUSTRALIAN SIGNPOST MATHS 1* • ISBN 978 1 4886 2178 9

5C One Half

CONCEPT

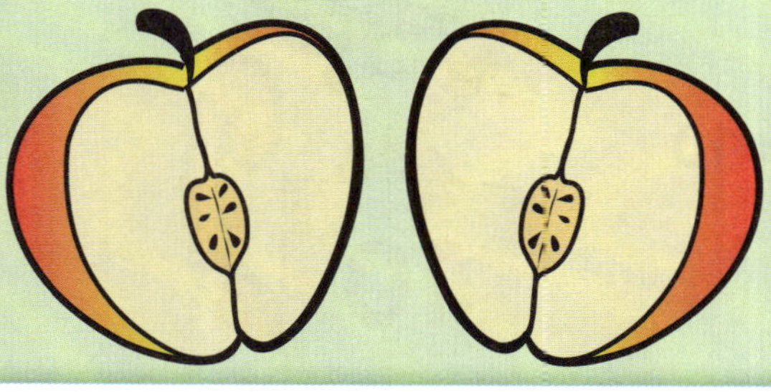

One half is one of two equal parts.

1 Circle one half of each.

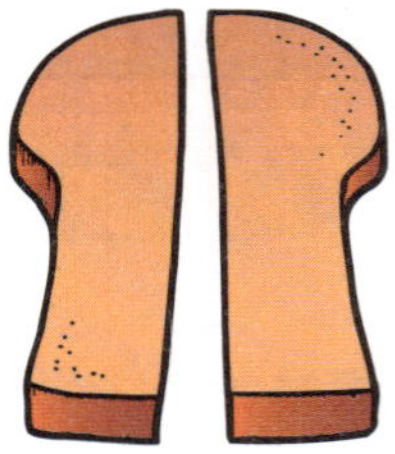

2 Draw a line to cut each object into halves. Colour one half.

INVESTIGATION

3 Use square paper to find three different ways to fold a square into halves. On each square below, draw a dotted line to show the fold, then colour one half. Discuss.

 • *AUSTRALIAN SIGNPOST MATHS 1* • ISBN 978 1 4886 2178 9

5D Introducing Graphs

1 a How many balls did we win?

b How many bears did we win?

c Which prize did we win most often?

d How many prizes did we win?

2 a Which room had most shoes?

b Which room had least shoes?

c How many shoes were in Room 3?

d How many shoes altogether?

3 a For how many days did Greg draw pictures?

b How many days were sunny?

c How many days were not sunny?

d Write a question of your own.

6A Numbers to 20

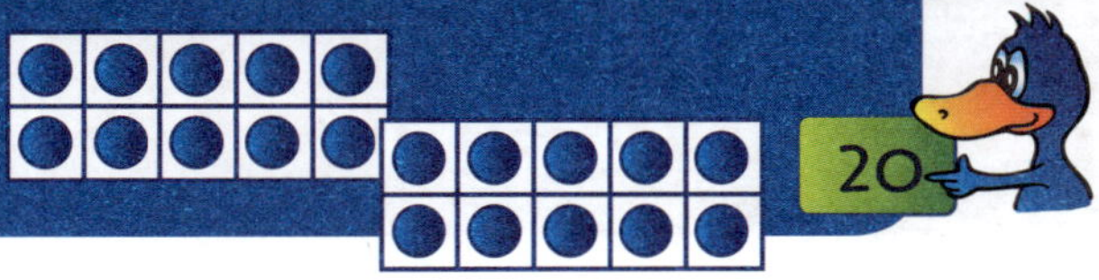

1 Write the missing numbers. Count forwards and backwards to 20.

1	2	3				7			
	12			15			18		20

2 Match each picture to a numeral. Circle groups of ten.

17

19

20

18

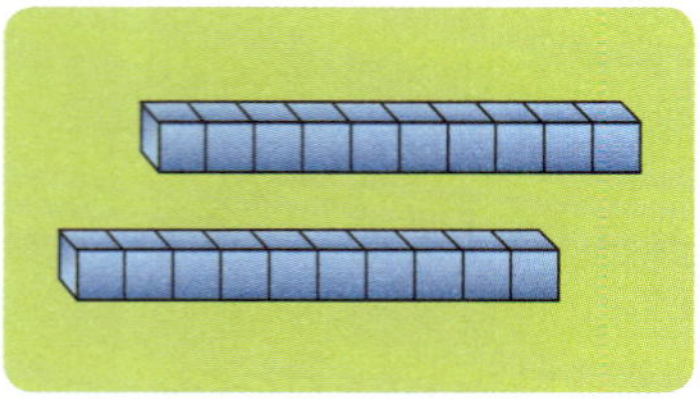

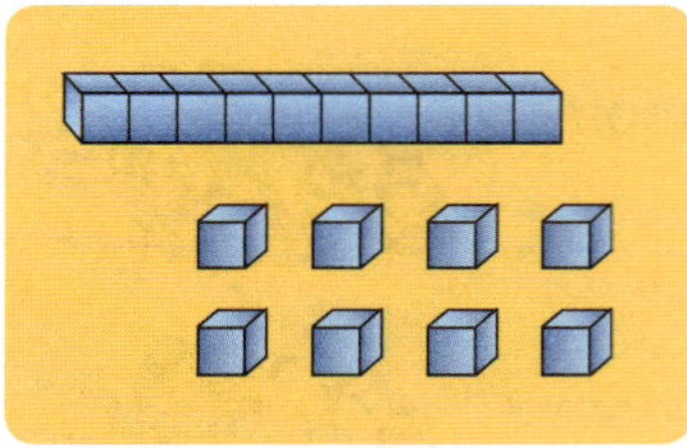

3 Write the numbers before and after.

a		10		b		13	
c		14		d		17	
e		18		f		19	

4 Two spiders. How many legs altogether?

ACTIVITY

6B Subtraction

CONCEPT

– means **take away**, **minus** or **subtract**.

❶ Complete:

a 6 fish take away 2 fish equals ☐.

☐ fish – ☐ fish = ☐ fish

☐ – ☐ = ☐

b

8 – 4 = ☐

c

3 – 3 = ☐

d

7 – 5 = ☐

e

9 – 6 = ☐

f

5 – 0 = ☐

g

10 – 2 = ☐

 • *AUSTRALIAN SIGNPOST MATHS 1* • ISBN 978 1 4886 2178 9

6C Patterns

CONCEPT

The number pattern matches the objects.

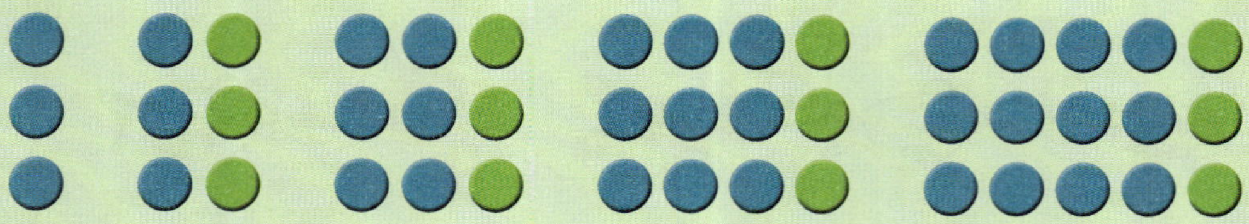

3 6 9 12 15

1 Complete each number pattern.

a

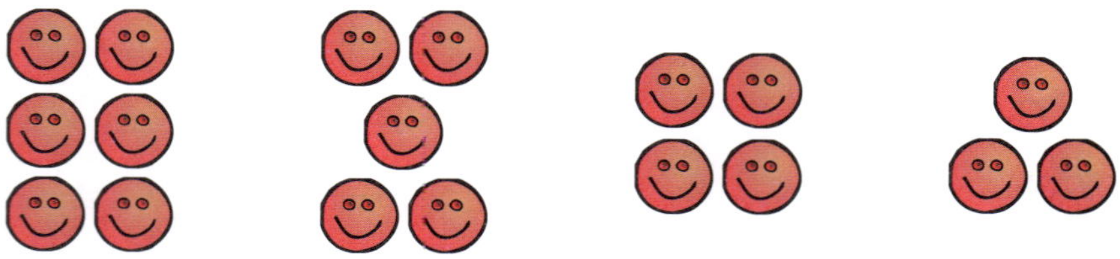

☐, ☐, ☐, ☐,

Subtract ☐ is the rule.

b

☐, ☐, ☐, ☐, ☐

Add ☐ is the rule.

c

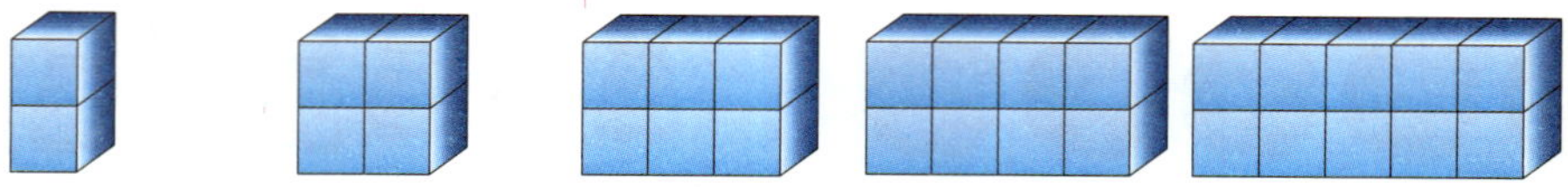

☐, ☐, ☐, ☐, ☐

Add ☐ is the rule.

ACTIVITY

Use blocks to make your own number patterns.

☐, ☐, ☐, ☐, ☐, ☐, ☐

6D Digital Time

analogue time

← 2 o'clock

digital time

← 2 o'clock

1 Write the time shown.

a 

☐ o'clock

b 

☐ o'clock

c 

☐ o'clock

d 

☐ o'clock

e 

☐ o'clock

f 

☐ o'clock

g 

☐ o'clock

h 

☐ o'clock

2 Show the analogue time for:

a 3 o'clock

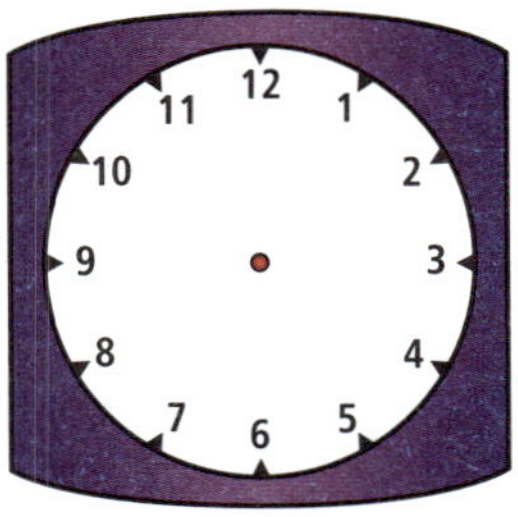

b 7 o'clock

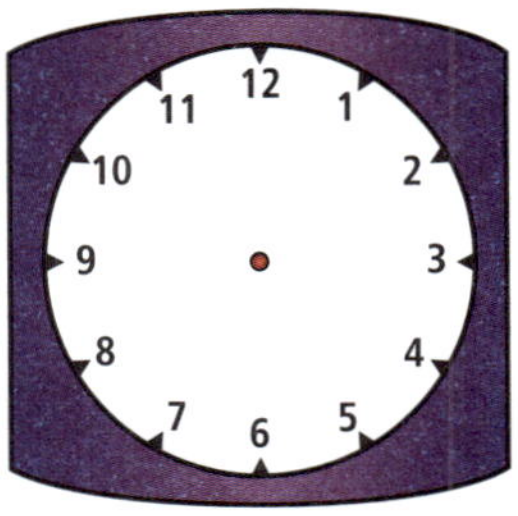

c 1 o'clock

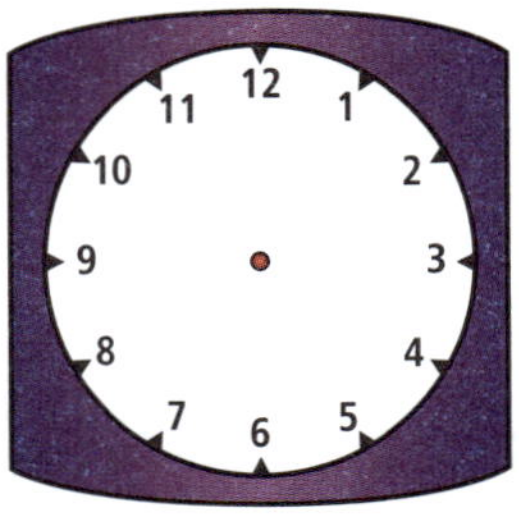

d 10 o'clock

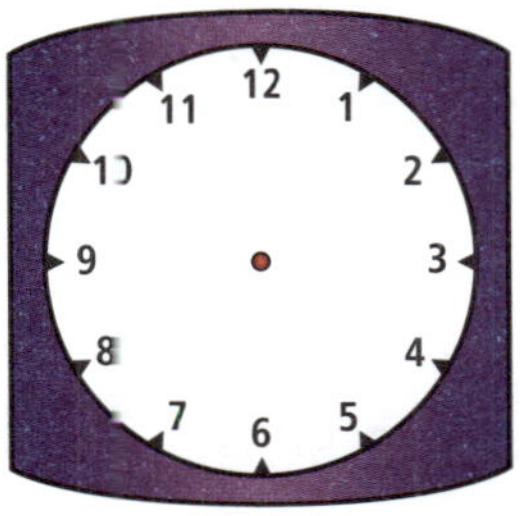

e 11 o'clock

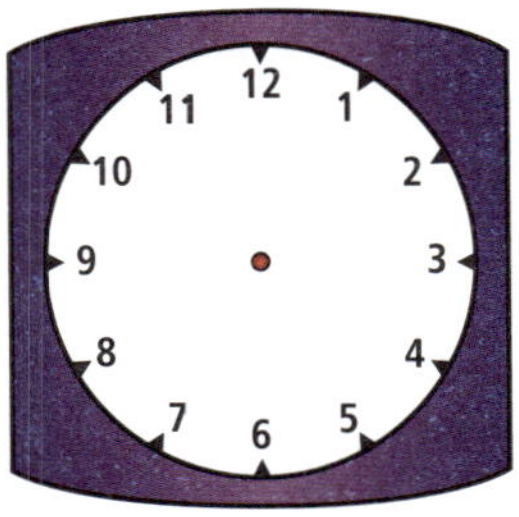

f 5 o'clock

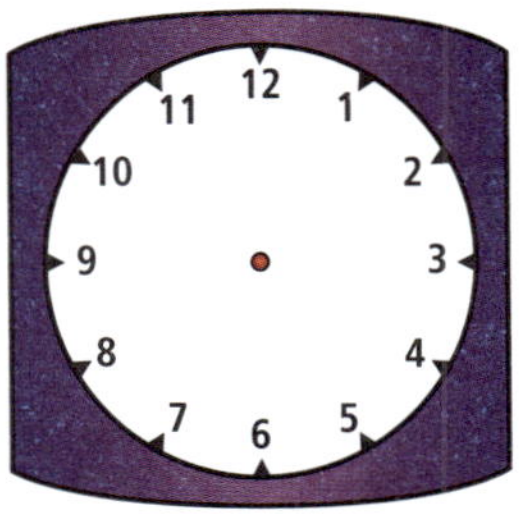

g 9 o'clock

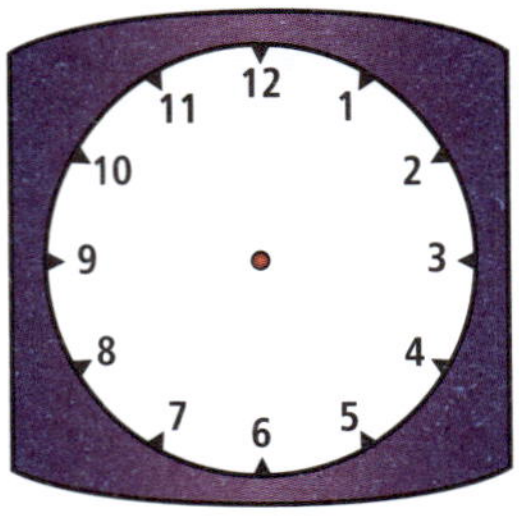

h 2 o'clock

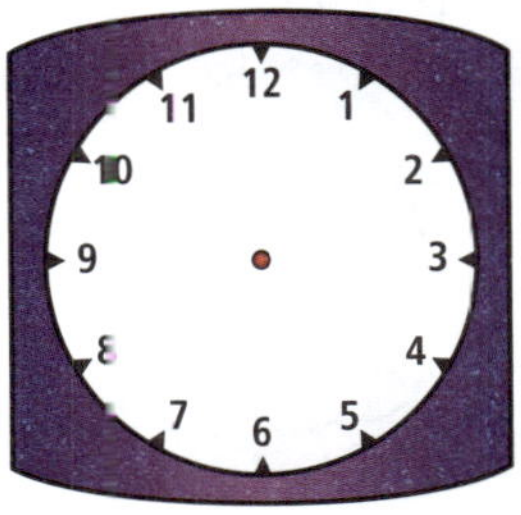

 • *AUSTRALIAN SIGNPOST MATHS 1* • ISBN 978 1 4886 2178 9

7A Partitioning to 10

1 Numbers can make patterns. Use the faces to find answers. Talk about patterns you can see.

1 and 8 makes ☐.

2 and 7 makes ☐.

3 and 6 makes ☐.

4 and 5 makes ☐.

5 and 4 makes ☐.

6 and 3 makes ☐.

7 and 2 makes ☐.

8 and 1 makes ☐.

1 + 9 = ☐

2 + 8 = ☐

3 + 7 = ☐

4 + 6 = ☐

5 + 5 = ☐

6 + 4 = ☐

7 + 3 = ☐

8 + 2 = ☐

9 + 1 = ☐

Use counters to make up some partitioning patterns of your own for 8, 7, 6, and 5.

4 = 1 + 3 →

4 = 2 + 2 →

4 = 3 + 1 →

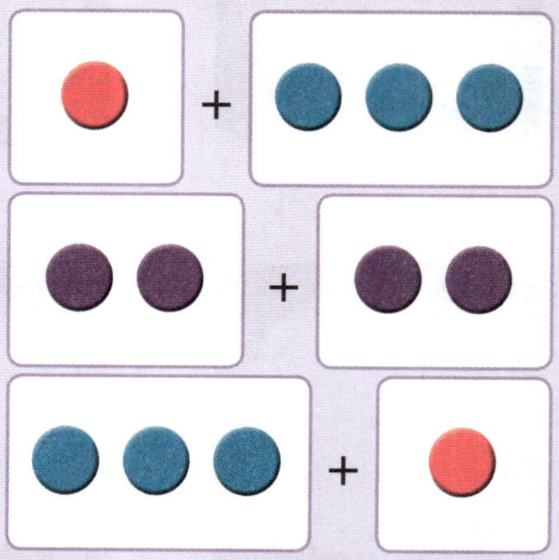

7B Sorting and Classifying Coins

CONCEPT

1 Discuss these coins. Colour the gold coins yellow and the silver coins green.

How many of each coin are there?

5c		10c		20c	
50c		$1		$2	

2 How many coins in each row?

a 

b 

7C Shapes in Our World

1 Match each photo with one of the objects in the middle.

2 Draw a simple shape that is in each object.

Use Plasticine to make models of these objects.

 • *AUSTRALIAN SIGNPOST MATHS 1* • ISBN 978 1 4886 2178 9

7D Using Pictures in Graphs

1

a How many people had straight hair? ☐

b How many people had curly hair? ☐

c How many people are shown altogether? ☐

d How many more people had straight hair than curly hair? ☐

2

a Which day was chosen least? ☐

b How many of the class chose Monday? ☐

c How many people chose a day? ☐

d What is your favourite day? ☐

8A Addition Facts

1 Complete:

a

☐ + ☐ = ☐

b

☐ + ☐ = ☐

2

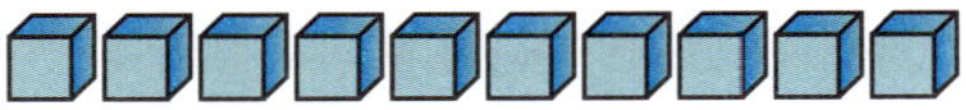

a 6 + 2 = ☐

b 8 + 0 = ☐

c 3 + 7 = ☐

d 0 + 10 = ☐

e 4 + 5 = ☐

f 1 + 6 = ☐

3 Match:

4 + 3	5	3 + 2
2 + 3	6	3 + 4
5 + 1	7	1 + 5
8 + 2	8	7 + 2
2 + 7	9	3 + 5
5 + 3	10	2 + 8

4 4 dogs, 3 cats, 2 mice. How many animals?

☐ + ☐ + ☐ = ☐

 • *AUSTRALIAN SIGNPOST MATHS 1* • ISBN 978 1 4886 2178 9

8B Doubles

When we add two numbers or groups that are the same, they are called **doubles**.

- Some doubles are: 1 + 1 | 2 + 2 | 3 + 3
- We say "Double 1", "Double 2", "Double 3".

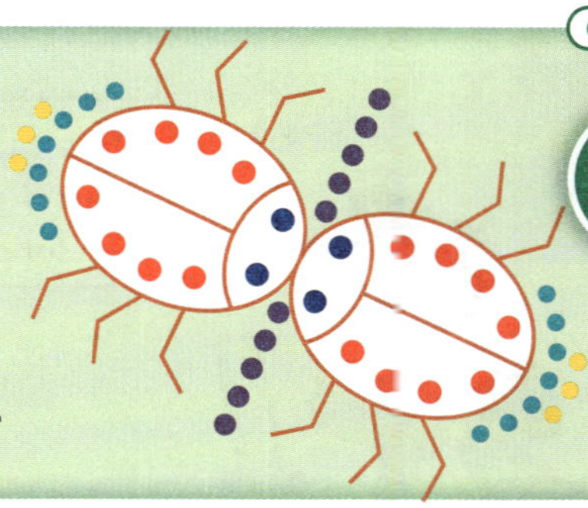

1 Use these pictures to complete the doubles.

a

2 horns and 2 horns

2 + ☐ = ☐

Double ☐ = ☐

b

4 legs and 4 legs

4 + ☐ = ☐

Double ☐ = ☐

c

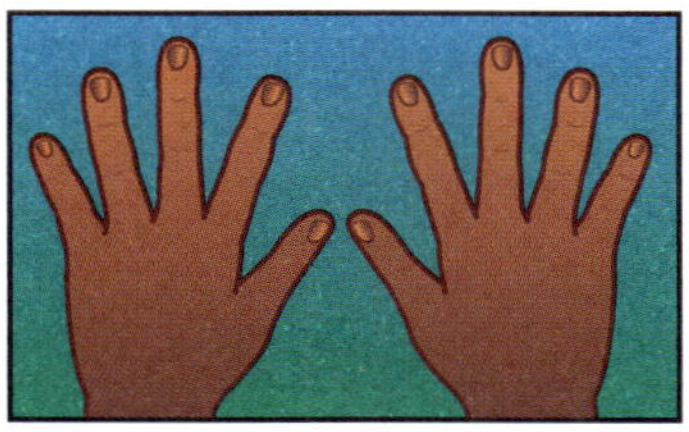

5 fingers and 5 fingers

☐ + ☐ = ☐

Double ☐ = ☐

2 List all the doubles you can find on the bugs at the top.

☐ + ☐ = ☐ ☐ + ☐ = ☐

☐ + ☐ = ☐ ☐ + ☐ = ☐

☐ + ☐ = ☐ ☐ + ☐ = ☐

8C Units of Length

CONCEPT

No gaps. No overlaps.

ACTIVITY

1 Use objects of the same size to measure the length of your desk.

Object	Length of Desk
This book	☐ books
A pen	☐ pens
A tens block	☐ blocks

2 Use your shoe to measure the width of each object.

☐

☐

☐

- Which is longest?

☐

- Which is shortest?

☐

INVESTIGATION

Count steps to measure distance.

- From your seat to the door: ☐ steps
- Along a path: ☐ steps
- ☐ ☐ steps.

 • *AUSTRALIAN SIGNPOST MATHS 1* • ISBN 978 1 4886 2178 9

8D Informal Units of Length

INVESTIGATION

1 Use pencils 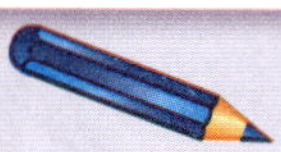 to measure the length:

a

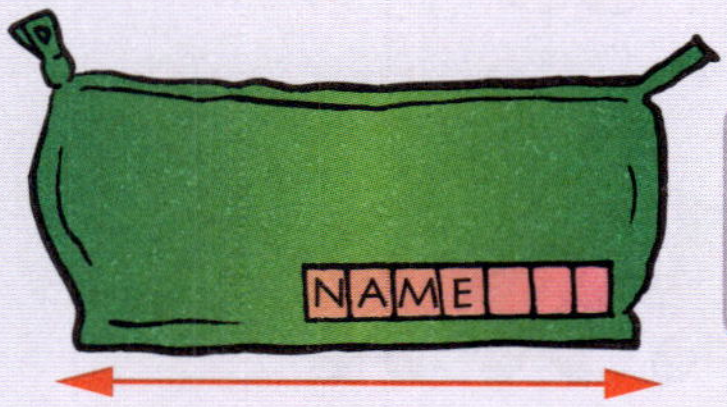

☐ pencils

b

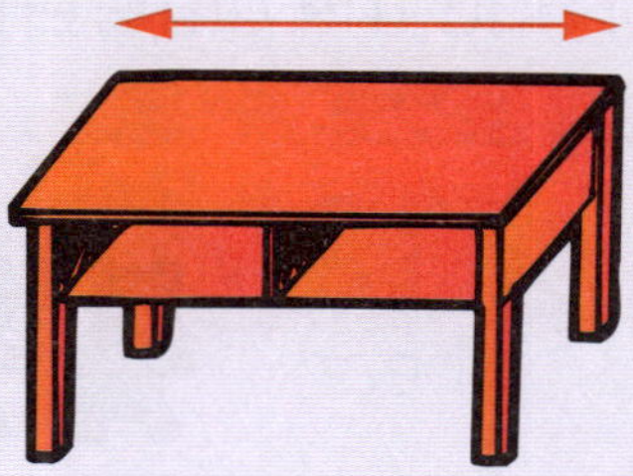

☐ pencils

c

☐ pencils

d

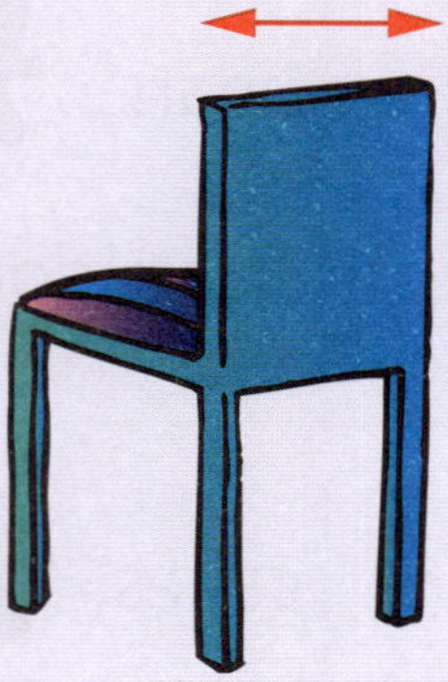

☐ pencils

2

Unit of Length	Length of Desk		Width of Desk	
	Guess	Check	Guess	Check

Why do we need more pegs than brushes?

ICT

Use pictures on a computer to measure the length of a line.

- This line is 5 platypuses long.

 • *AUSTRALIAN SIGNPOST MATHS 1* • ISBN 978 1 4886 2178 9

9A Difference

– means the **difference between.**
You can find the difference by counting on.

1 Sam has bears.

Jess has bears.

How many more bears does Sam have?

$8 - 4 =$

2 Lisa has cakes.

Tom has cakes.

How many more cakes does Lisa have?

$9 - 5 =$

3 Tina has bells.

Ben has bells.

How many more bells does Tina have?

$10 - 6 =$

4 Tim has bananas.

Sara has bananas.

How many more bananas does Tim have?

– =

 ISBN 978 1 4886 2178 9

9B Difference

– means the **difference between**.
You can find the difference by counting on.

CONCEPT

1. How many are in each group? Answer the questions.

a

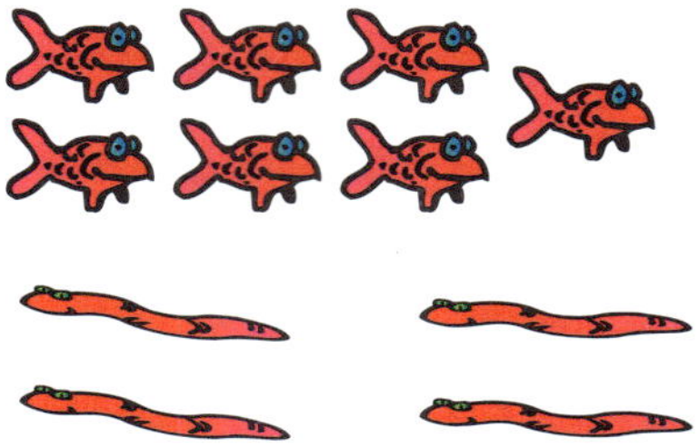

How many more fish than worms?

$7 - 4 =$

b

How many more buckets than spades?

$\square - \square = \square$

c Zoe read .

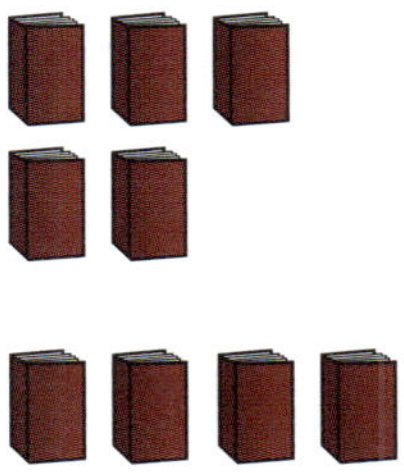

Tom read .

How many more books did Zoe read?

$\square - \square = \square$

d Jake ate .

Mina ate .

How many more did Jake eat?

$\square - \square = \square$

ACTIVITY

Make 2 lines of objects and find the difference.
Write a number sentence to match.

$\square - \square = \square$

 • *AUSTRALIAN SIGNPOST MATHS 1* • ISBN 978 1 4886 2178 9

9C Comparing Mass

Hefting is used to compare mass.

1 Compare the masses of these objects by hefting.

a

Pencil Book

The ______ is heavier.

The ______ is lighter.

b

Shoe Box

The ______ is heavier.

The ______ is lighter.

c

Clock Cup

The ______ is heavier.

The ______ is lighter.

2 Compare the masses of different objects by hefting. Use balance scales to check.

My heaviest object was ______.

 • *AUSTRALIAN SIGNPOST MATHS 1* • ISBN 978 1 4886 2178 9

9D The Oval and Other Shapes

CONCEPT

An oval is like a squashed circle. It has one curved side.

1 Draw 4 ovals, 3 squares, 1 triangle and 2 rectangles.

2 What shapes were used to make this fish?

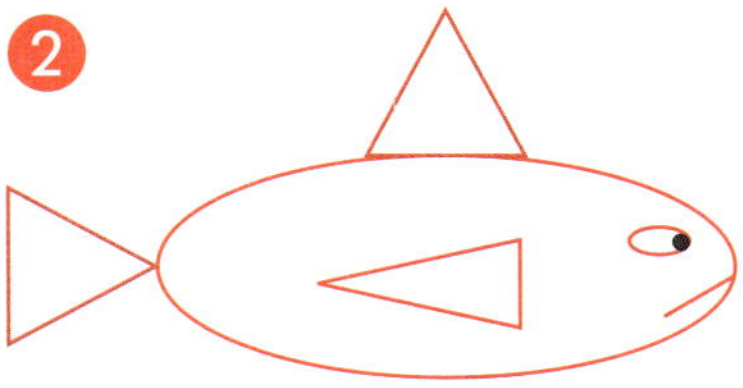

Choose two shapes and make your own picture.

3 How many of these shapes do you see?

triangles

squares

rectangles

ovals

10A Groups of 10

We can put large numbers into groups of 10.

This is 4 groups of ten.
There are 40 pencils.

1 Count the groups of ten.

a  ☐ tens

b 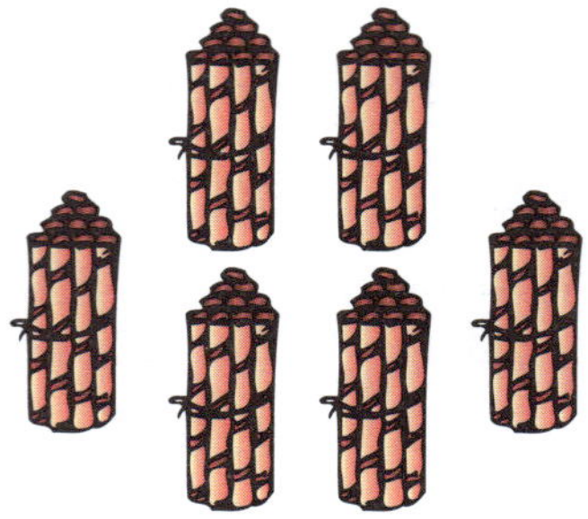 ☐ tens

c  ☐ tens

d  ☐ tens

e 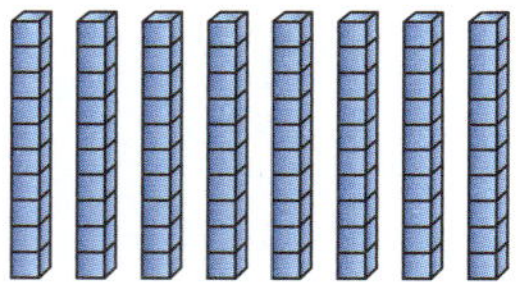 ☐ tens

f 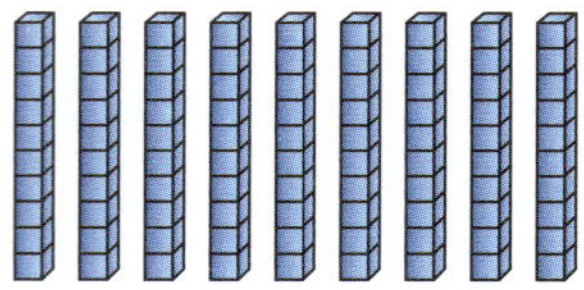 ☐ tens

2 Circle groups of ten.

☐ tens

Talk about your answer.

3

☐ tens

☐ tens

3 tens is written as 30.

Tens
3

= 30

1 Circle groups of 10 and complete the number sentences.

a

Tens [ ] = [ ]

b

Tens [ ] = [ ]

c

Tens [ ] = [ ]

2 Help Joey jump across the river. Jump by tens from 10 to 90.

 • *AUSTRALIAN SIGNPOST MATHS 1* • ISBN 978 1 4886 2178 9

10C Australian Coins

1 Discuss these coins. Colour the gold coins yellow and the silver coins green.

How many coins? 5c ☐ 10c ☐ 20c ☐

50c ☐ $1 ☐ $2 ☐

2 a Discuss the value of the coins in Question 1.

b Name the animals on the 5c, 10c, 20c, 50c and $1 coins.

c Whose head is shown on the back of each coin?

 • *AUSTRALIAN SIGNPOST MATHS 1* • ISBN 978 1 4886 2178 9

10D Left and Right

CONCEPT

Do you know left and right?

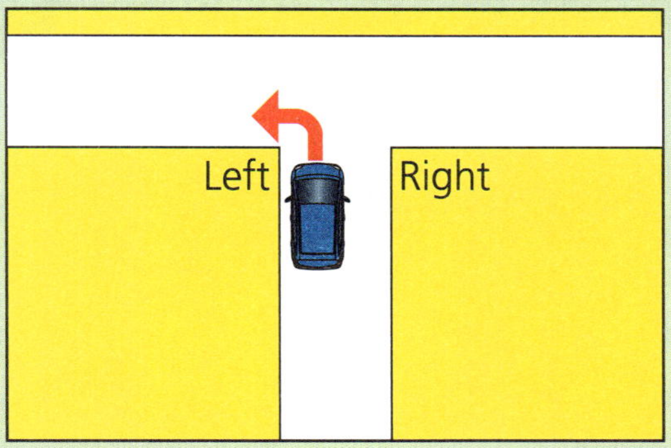

The blue car is turning left.

❶ Show how each car will turn.

a Turn right.

b Turn left.

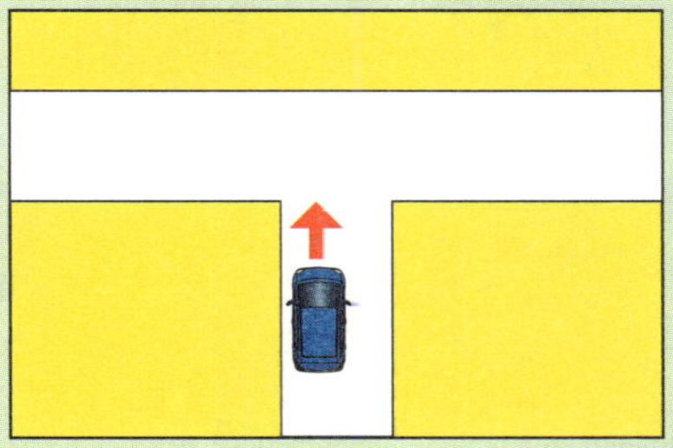

❷ Which are left and which are right?

a

The striped monkey is on the [ ].

The spotted monkey is on the [ ].

b

The bird is on the [ ].

The ball is on the [ ].

The pants pocket is on the [ ].

c

The [ ] is on the left.

11A Subtraction

CONCEPT

– means **take away** or **minus**.

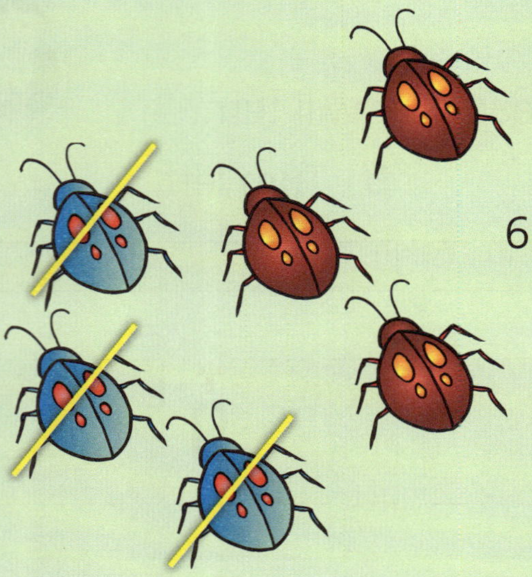

6 bugs take away 3 bugs equals ☐.

☐ bugs – ☐ bugs = ☐ bugs

6 ☐ 3 = ☐

1

11 – 4 = ☐

2

10 – 5 = ☐

3

15 – 10 = ☐

4

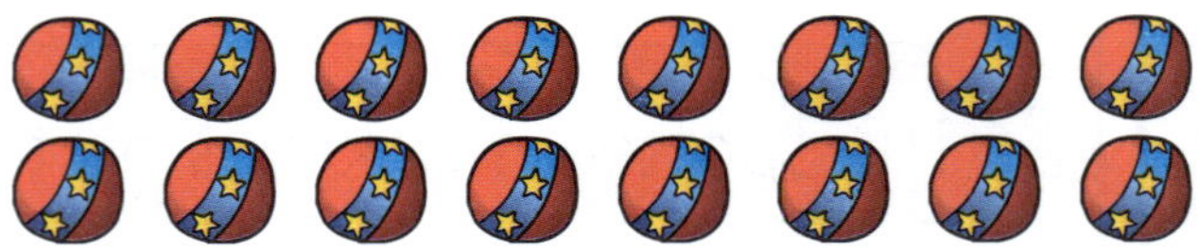

16 – 0 = ☐

5

18 – 6 = ☐

6

20 – 10 = ☐

11B Difference Between Groups

– also means the **difference between**.
Find the difference by counting on.

CONCEPT

1 

____ – ____ = ____

There are ____ more  than . The difference is ____.

2 

____ – ____ = ____

There are ____ more  than . The difference is ____.

3 

____ – ____ = ____

There are ____ more  than . The difference is ____.

4 

____ – ____ = ____

There are ____ more  than . The difference is ____.

ACTIVITY

Make 2 lines of objects and find the difference.

Write a number sentence to match.

 ISBN 978 1 4886 2178 9

11C Position Language

CONCEPT

- The triangle is on the left.
- The square is on the right.
- The circle is between the triangle and the square.

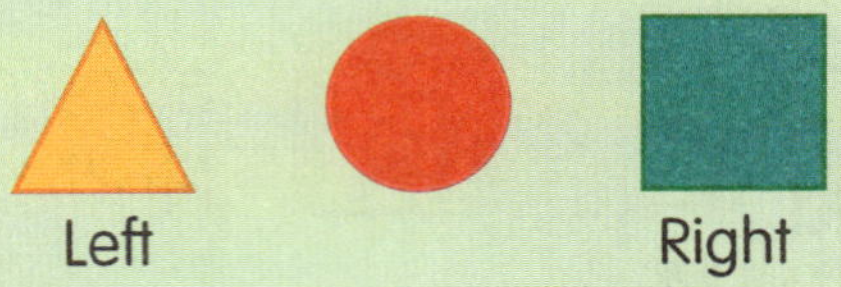

1. Use the picture to answer these questions. Choose the correct position word.

Position Words	
next to	towards
above	between
left	right
in front of	

a The bucket is ______ the sandcastle.

b The cloud is ______ the water.

c The girl is on the ______ of the picture.

d The flag is on the ______ of the picture.

e The girl is walking ______ the flag.

f The sandcastle is ______ the bucket and the flag.

2.

a The boy is on the ______.

b The bird is on the ______.

 • *AUSTRALIAN SIGNPOST MATHS 1* • ISBN 978 1 4886 2178 9

11D Location Language

1 Look at the picture and circle the correct words in each sentence.

a The caterpillar is crawling along / beneath the stick.

b The mouse is above / below the hawk.

c The house is between / inside the hills.

d The hills are far from / close to the tree.

e The car is in front of / behind the house.

f The ant is under / over the tree.

g The kookaburra is perched on / over the branch.

h The girl is to the right / left of the pram.

i The bat is hanging upside down / right side up.

2 Where would the girl go to take the apples into the house?

→ Talk about this.

3 In this picture, draw:

a a fork on the left of the plate

b a knife on the right of the plate

c a star in the centre of the window

d a book on a corner of the table

e a bottle on the edge of the table

f a picture above the clock

Colour the picture.

 • *AUSTRALIAN SIGNPOST MATHS 1* • ISBN 978 1 4886 2178 9

12A Groups of 10

CONCEPT

ten	10
twenty	20
thirty	30
forty	40
fifty	50
sixty	60
seventy	70
eighty	80
ninety	90

1 Write the numerals.

a 

1 ten = ___

b 

3 tens = ___

c 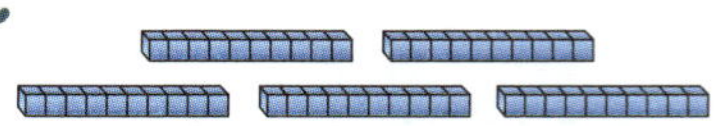

5 tens = ___

d 

2 tens = ___

e 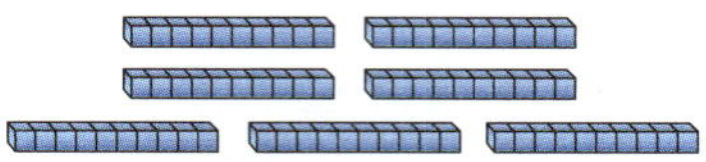

7 tens = ___

f 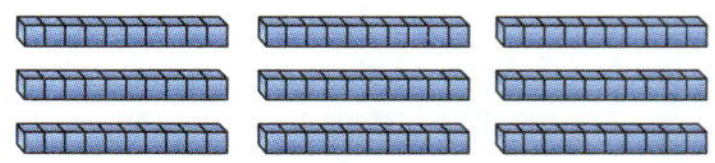

9 tens = ___

ACTIVITY

- Model these numbers with place-value blocks.
  - 10
  - 40
  - 30
  - 80
  - 90
  - 70

ICT

- Make these numbers on a calculator.
  - 10
  - 40
  - 30
  - 80
  - 90
  - 70

 ISBN 978 1 4886 2178 9

12B Counting by 10s

10	20	30	40	50
ten	twenty	thirty	forty	fifty
60	70	80	90	100
sixty	seventy	eighty	ninety	one hundred

1 a Count forwards by tens to 100.

b Count backwards by tens from 100.

2 Write these numbers as numerals.

a twenty ☐ b eighty ☐ c ninety ☐ d forty ☐

3 Write the next four numbers in each pattern.

a 10, 20, 30, ☐, ☐, ☐, ☐

b 90, 80, 70, ☐, ☐, ☐, ☐

4 Begin at 10 and join the dots to count forwards by tens.

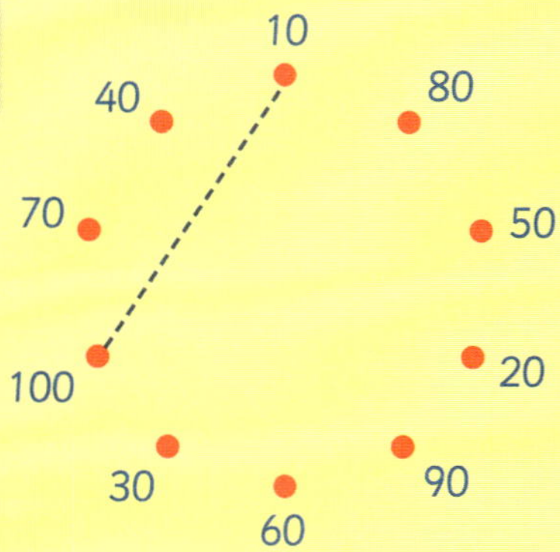

5 How much money?

☐ cents

12C Difference

– means the **difference between**.

You can find the difference by counting on.

CONCEPT

1 Jane has ☐ ducks.

Alex has ☐ ducks.

☐ has ☐ more ducks than ☐.

2

☐ – ☐ = ☐

There are ☐ more  than . The difference is ☐.

3 How many more caps has Aki than Ella?

- Aki's caps: 
- Ella's caps: 

☐ – ☐ = ☐

Aki has ☐ more caps.

ACTIVITY

4 How many more dots than ☐?

Dots ●●●●●●●●●●●●●●●●●

Crosses XXXXXXXXXXXXXXX

☐ – ☐ = ☐

 ISBN 978 1 4886 2178 9

12D The Hexagon

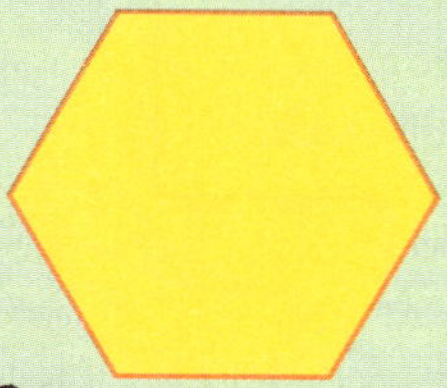

The hexagon has **6 sides** and **6 corners**.

Shape Words

Circle Oval Triangle Square Rectangle

1. Name the shapes used.

2. Trace each shape. Write its name.

3. Colour the hexagons red.

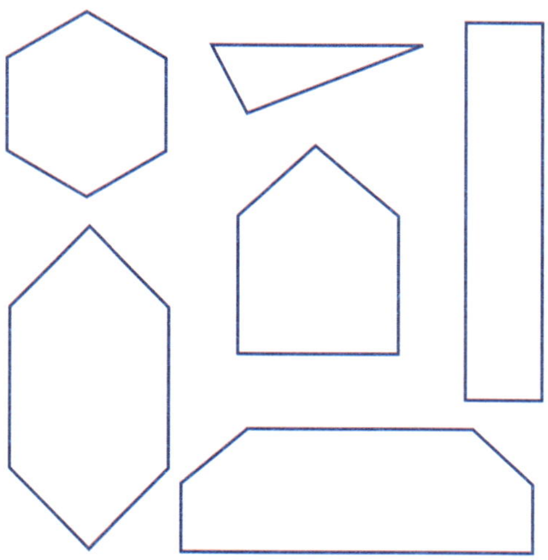

4. Draw a hexagon.

5. A hexagon has ☐ sides and ☐ corners.

© PEARSON AUSTRALIA 2018 • *AUSTRALIAN SIGNPOST MATHS 1* • ISBN 978 1 4886 2178 9

13A Counting On

CONCEPT

●●●●●	●	●	●
5	6	7	8

5 and 3 more = 8

1 Count on to find how many altogether.

a

●●●	●	●
3	4	5

3 and 2 more = ☐

b

●●●●●●	●	●
6	7	8

6 and 2 more = ☐

c

●●●●●●	●	●	●	●
6	7	8	9	10

6 and 4 more = ☐

d

●●●●●●●●	●
8	9

8 and 1 more = ☐

e

10	●	●	●	●

10 and 4 more = ☐

f

13	●	●	●

13 and 3 more = ☐

ACTIVITY

Use counters to count on from a chosen number.
Discuss and record your number sentences.

13B Shape Hunt

1 Draw things that match each 3D object.

ball-shaped object

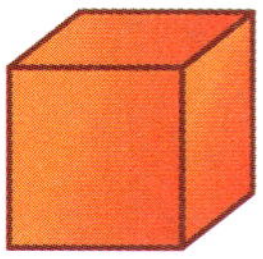

box-shaped object

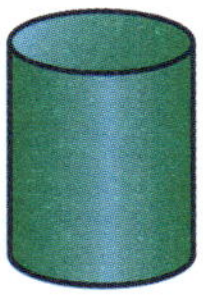

can-shaped object

cone-shaped object

ACTIVITY

Use Plasticine to make a model of each object.

 • *AUSTRALIAN SIGNPOST MATHS 1* • ISBN 978 1 4886 2178 9

13C Time Revision

This is 4 o'clock.

1 Write the time.

a

☐ o'clock

b

☐ o'clock

c

☐ o'clock

3 o'clock → **3:00**

2 Read and write each time.

a

☐ o'clock

b

☐ o'clock

c

☐ o'clock

3 Complete:

a 4 o'clock =

b

= ☐ o'clock

 • *AUSTRALIAN SIGNPOST MATHS 1* • ISBN 978 1 4886 2178 9

13D Analogue and Digital Time

1 Read each time. Complete the labels.

a

☐ o'clock

b

☐ o'clock

c

☐ o'clock

d

☐ o'clock

2 Read the times and complete the sentences. Write the same time on the digital clock.

a I get up at ☐ o'clock.
☐ : ☐

b Breakfast is at ☐ o'clock.
☐ : ☐

c The bell rings at ☐ o'clock.
☐ : ☐

d Recess time is at ☐ o'clock.
☐ : ☐

e The bell rings at ☐ o'clock.
☐ : ☐

f Dinner is at ☐ o'clock.
☐ : ☐

3

This clock shows ☐ time.

14A Numbers to 100

Numbers can be written in groups of **tens** and **ones**.

2 tens 4 ones

24

4 tens 2 ones

42

1 Write the number modelled.

a

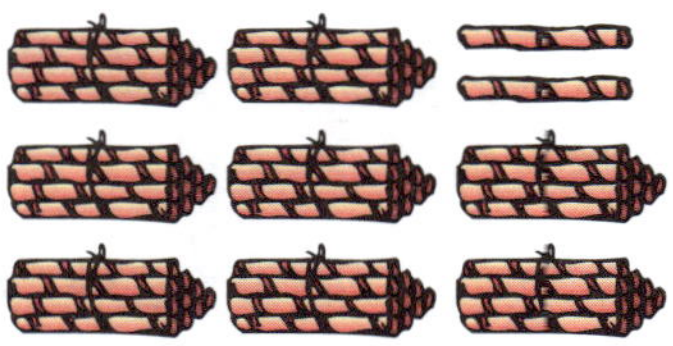

tens | ones

b

tens | ones

c

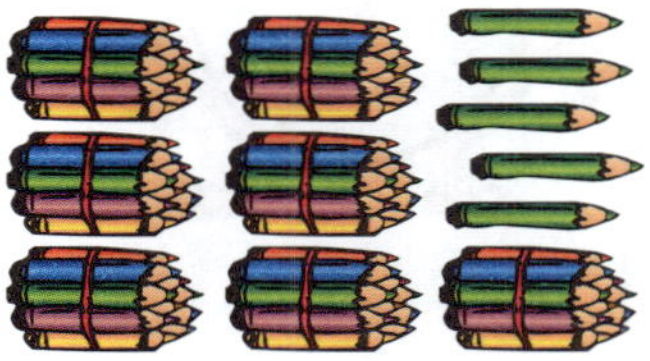

tens | ones

2 From Question 1, write the number that is:

a the smallest

b the biggest

3 Using the beads around the page, take turns to show where you would find these numbers.

a 13 **b** 29 **c** 45 **d** 67 **e** 80 **f** 105

4 On a calculator, make the numbers above. For each number say how many tens and how many ones.

Start

 ISBN 978 1 4886 2178 9

14B Numbers to 100

1 For each, complete the numeral expander and label.

a
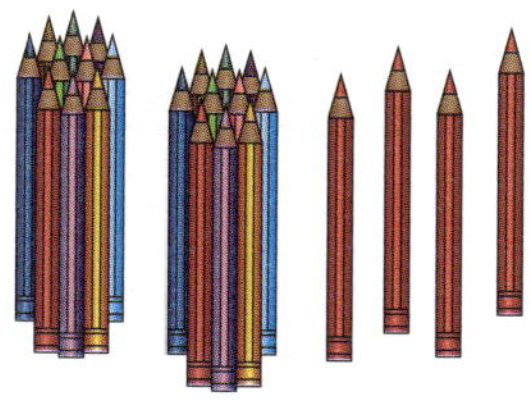

2	Tens	4	Ones

b

	Tens		Ones

c
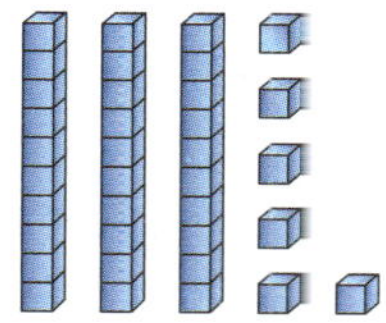

	Tens		Ones

d
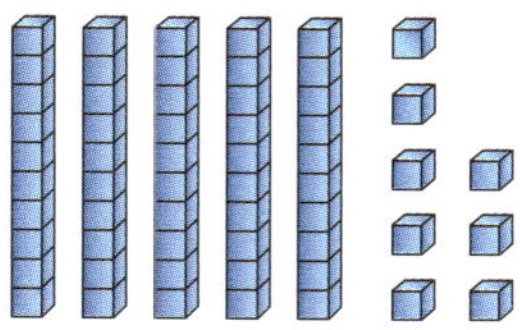

	Tens		Ones

e
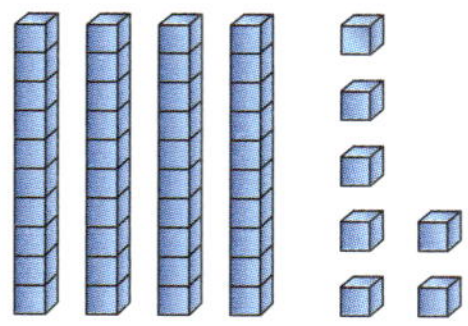

	Tens		Ones

f
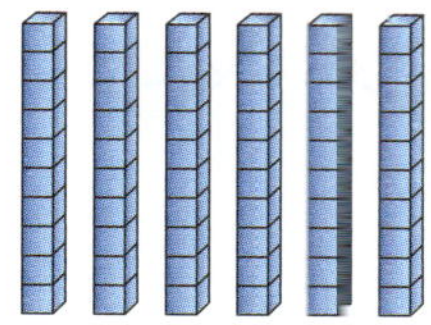

	Tens		Ones

g
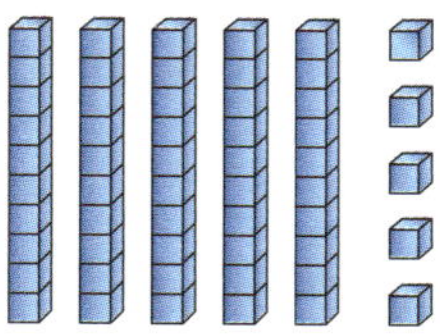

	Tens		Ones

h
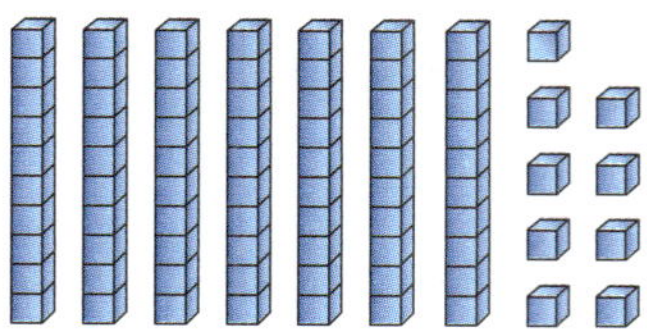

	Tens		Ones

i

	Tens		Ones

With a partner, use place-value blocks to model these numbers. Talk about your answers.

- 10
- 27
- 40
- 30
- 64
- 83
- 50
- 79
- 100
- 90

14C Counting Back

CONCEPT

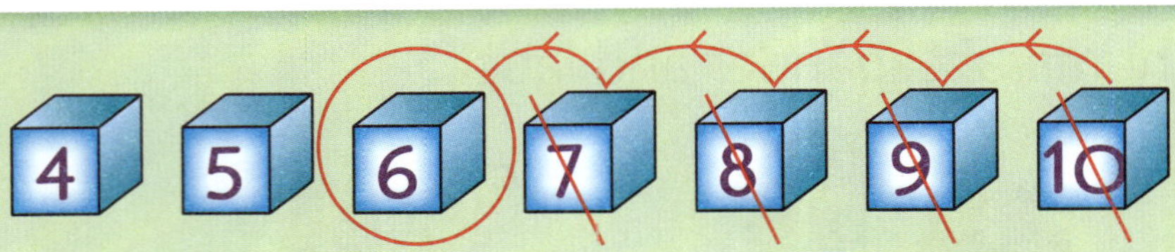

10 – 4 = 6

Count back 4 from 10. Cross them off as you count.

1 Count back to complete.

a

8 take away 1 leaves ☐.

b

11 take away 2 leaves ☐.

c

10 take away 3 leaves ☐.

d

8 take away 3 leaves ☐.

2 Count back to take away two.

7	☐	☐
10	☐	☐
12	☐	☐

3 Count back to take away three.

15	☐	☐	☐
18	☐	☐	☐
20	☐	☐	☐

 ISBN 978 1 4886 2178 9

14D Drawing Shapes

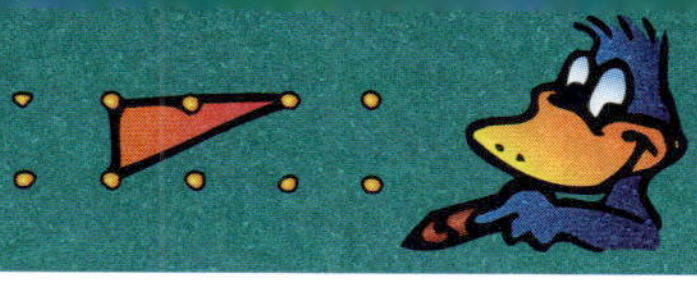

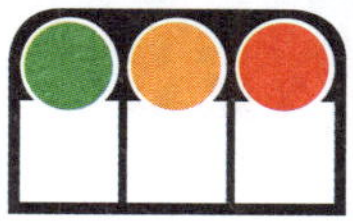

CONCEPT

Erica drew these shapes. **B** has 3 sides and 3 corners.

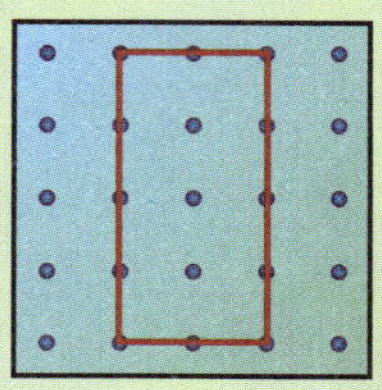

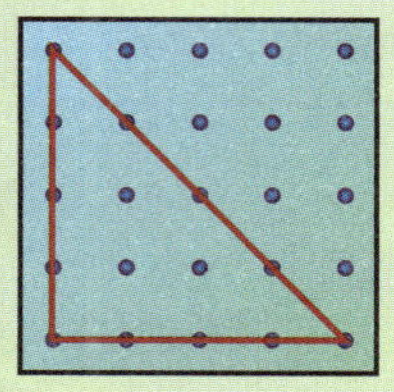

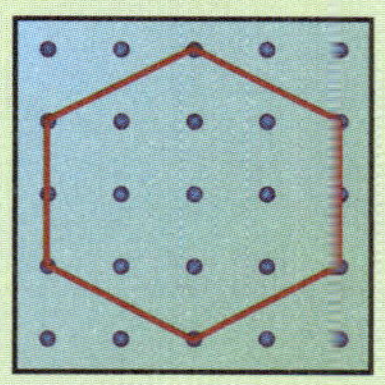

1 Draw:

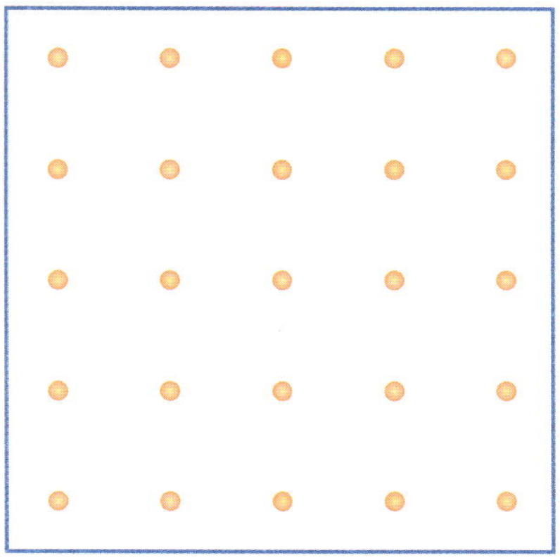

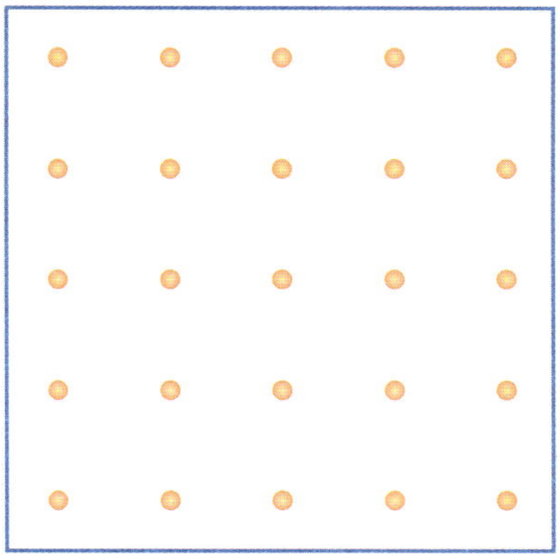

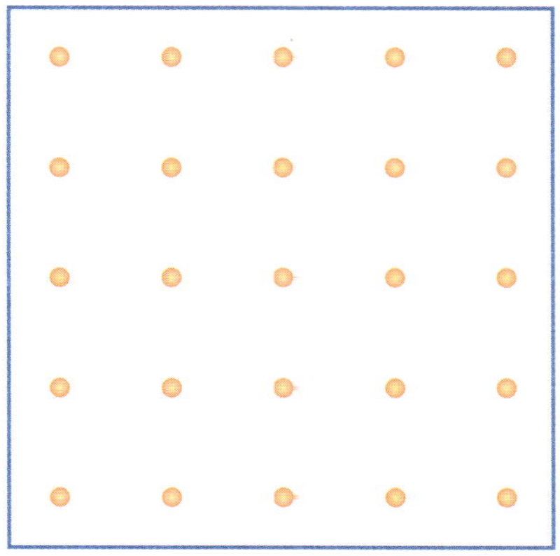

a a rectangle **b** a triangle **c** a square

2 Matt made these shapes.
Draw some shapes of your own.

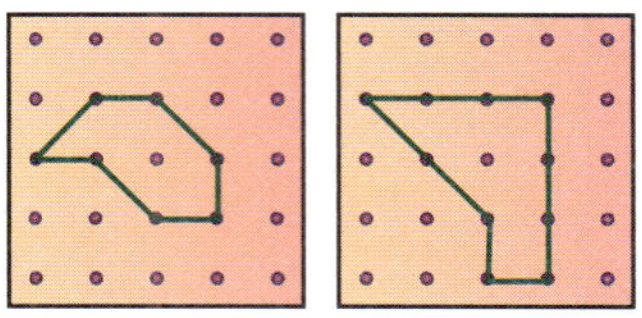

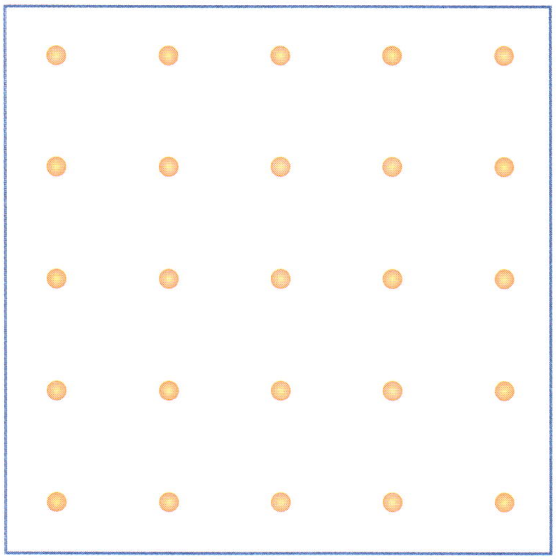

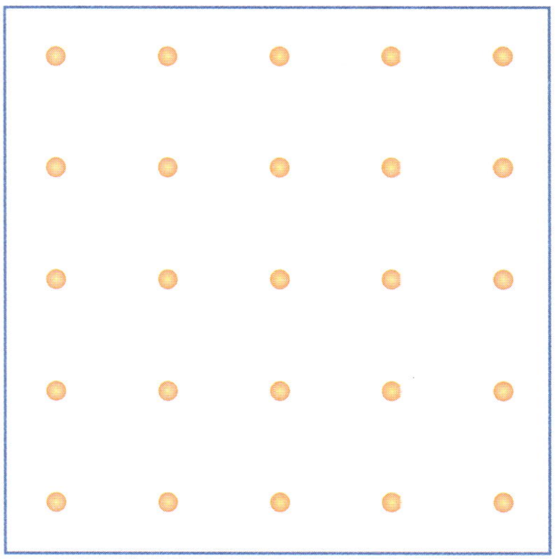

a How many corners? ☐

b How many corners? ☐

15A The Number Line

CONCEPT

0 1 2 3 4

1 and 4 are marked.

25 26 27 28 29

26 and 28 are marked.

1 Write the numbers that are marked.

a 0 1 2 3 4 5

b 11 12 13 14 15 16

c 20 21 22 23 24

d 40 45 50 55 60 65

e 40 50 60 70 80 90

2 Show each group of numbers on the number line.

a 1, 2, 3

0 1 2 3 4 5

b 13, 15, 17

12 13 14 15 16 17

c 26, 27, 28, 29

26 27 28 29 30 31

d 54, 55, 56, 58

53 54 55 56 57 58

e 74, 75, 77, 79

74 75 76 77 78 79

f 88, 90, 91, 92

87 88 89 90 91 92

g 70, 80, 90

50 60 70 80 90 100

h 85, 95, 100

75 80 85 90 95 100

15B The Calculator

CONCEPT

ON turns the calculator **on**.

OFF turns the calculator **off**.

C **clears** the calculator.

+ means **plus**.

= means **is equal to**.

1. Use a calculator to complete.

 plus 

a [8] [+] [2] [=] ☐

 plus 

b [7] [+] [9] [=] ☐

 plus 

c [4] [+] [3] [=] ☐

 plus 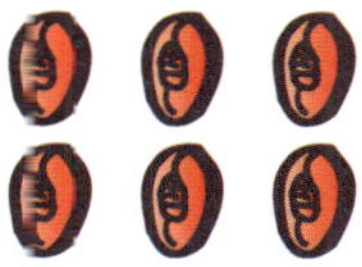

d [7] [+] [6] [=] ☐

ICT

2. Press the keys in this order and write the pattern you see.

[0] [+] [1] [=] [=] [=] [=] [=] [=]

The pattern is: ☐ ☐ ☐ ☐ ☐ ☐

15C Informal Units of Capacity

INVESTIGATION

1. How many fill the large container?

	Cups of Water		Bottles of Water	
pan 	Guess	Check	Guess	Check
ice-cream container 	Guess	Check	Guess	Check
jug 	Guess	Check	Guess	Check
margarine container 	Guess	Check	Guess	Check
bucket 	Guess	Check	Guess	Check

2. How many times can one jug fill each container?

Jug

a cup

b glass

c eggcup

15D Informal Units of Capacity

1 Circle the object that will hold the most. Colour the object that will hold the least.

Eggcup

Drink Can

INVESTIGATION

2 How many does each hold?

a Use matchboxes. 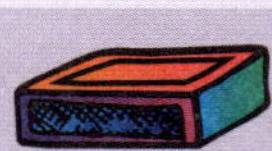

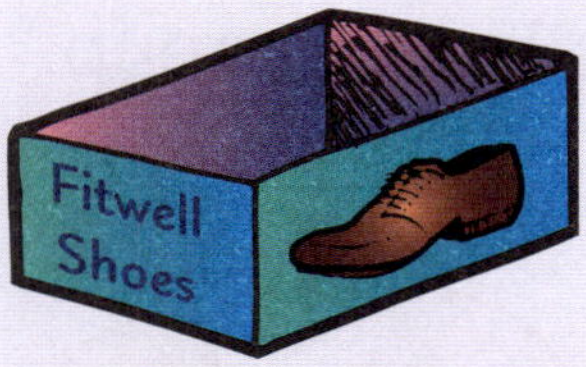

Guess Check

Guess Check

b Use marbles. 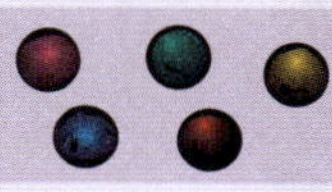

Guess Check

Guess Check

c Use place-value ones blocks. 

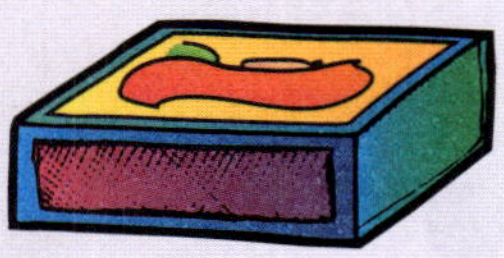

Guess Check

Guess Check

d Use place-value tens blocks. 

Guess Check

Guess Check

 • *AUSTRALIAN SIGNPOST MATHS 1* • ISBN 978 1 4886 2178 9

16A Numbers to 20

1 Write the numerals 11 to 19 in order. Colour each numeral and its name the same colour.

thirteen · fourteen · seventeen · eighteen

									20

twelve · fifteen · eleven · sixteen · nineteen

2 Guess how many marbles are in each bag. Circle the largest group.

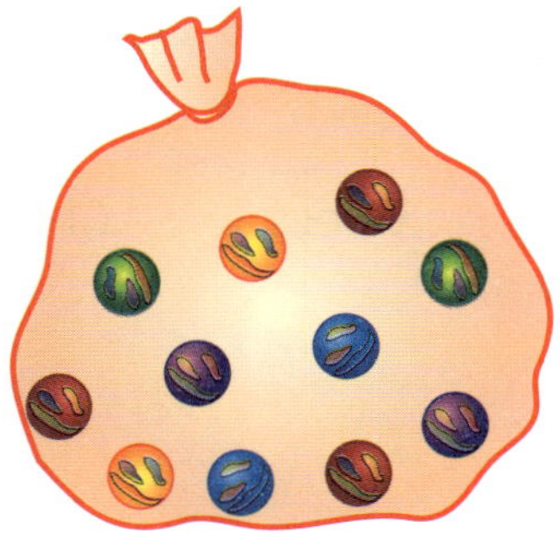

Guess Count

Guess Count

Guess Count

3 Order these numbers from smallest to largest.

16 19 11 13

4 Estimate and count the number of marbles.

Guess Count

Talk about your answer.

16B Addition to 20

+ means **add**, **and** or **plus**.

= means **makes** or **is equal to**.

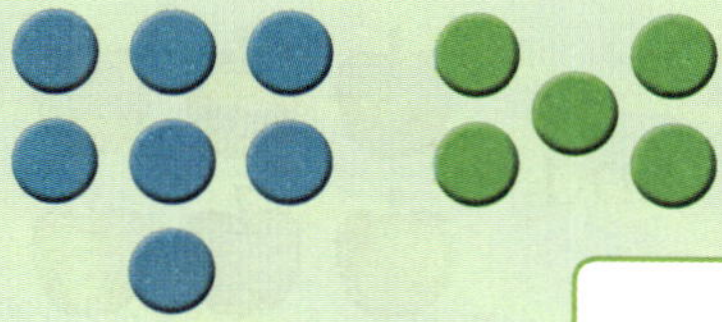

7 + 5 = ☐

1 Complete the number sentences.

a

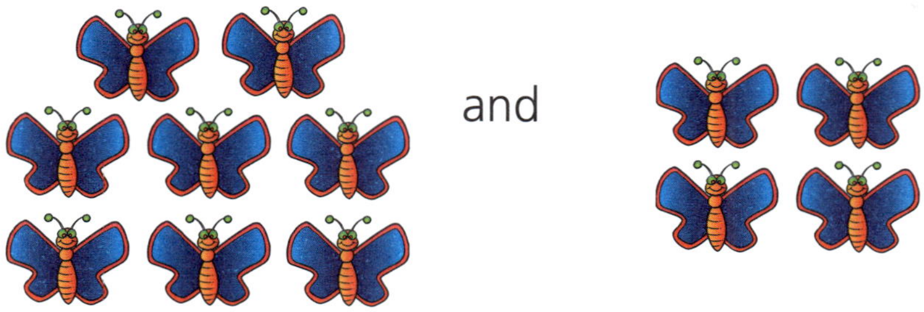

and

8 + 4 = ☐

b

and

9 + 7 = ☐

c

and

10 + 10 = ☐

Draw pictures to make up your own number stories. Write a number sentence to match.

☐ + ☐ = ☐

16C Addition to 20

1  and 

☐ + ☐ = ☐

2  and 

☐ + ☐ = ☐

3  and 

☐ + ☐ = ☐

4  and 

☐ + ☐ = ☐

5 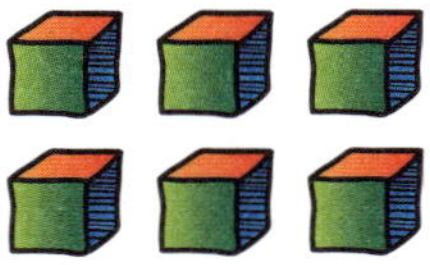 and 

☐ + ☐ = ☐

6  and 

☐ + ☐ = ☐

7  and 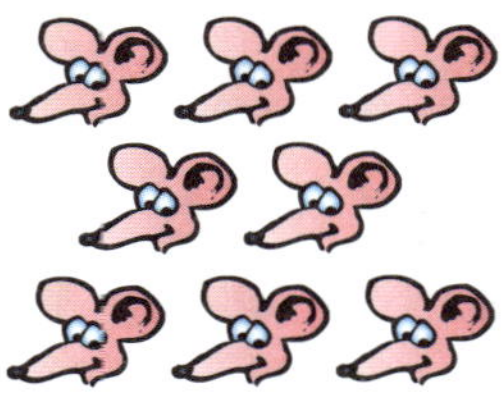

☐ + ☐ = ☐

8 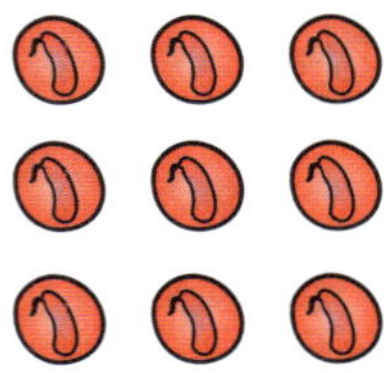 and 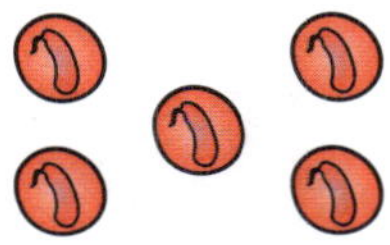

☐ + ☐ = ☐

INVESTIGATION

Work with a partner.

- Start with less than 20 counters. Give some to your partner.
- Make up a number story to add the two groups.
- Write a number sentence to go with the story.

16D Chance Words

FUN SPOT

Use these words to talk about the picture.

will happen | won't happen | might happen

17A Numbers to 100

ACTIVITY

1 On the hundred chart:

a Count by ones to 100.

b Count by twos to 50.

c Count by fives to 100.

d Count by tens to 100.

2 a Start at 20 and count back by ones.

b Start at 100 and count back by ones.

1	2	3	4	5	6	7	8	9	10
11	12	13	14	15	16	17	18	19	20
21	22	23	24	25	26	27	28	29	30
31	32	33	34	35	36	37	38	39	40
41	42	43	44	45	46	47	48	49	50
51	52	53	54	55	56	57	58	59	60
61	62	63	64	65	66	67	68	69	70
71	72	73	74	75	76	77	78	79	80
81	82	83	84	85	86	87	88	89	90
91	92	93	94	95	96	97	98	99	100

3 Write the numeral.

a twenty-five ☐

b forty-one ☐

c seventy-one ☐

d ninety ☐

e thirty-eight ☐

f eighty-three ☐

4 Write the numbers before and after.

a

Before		After
	69	
	37	
	41	

b

Before		After
	83	
	55	
	99	

17B Numbers to 100

1 Write the numeral.

a sixty-one

b twenty-two

c eighty-four

d seventy-nine

2 Write the number in words. Remember to use a hyphen.

a 65

b 43

c 80

d 27

e 79

3 Write the number that is 1 more than:

a 33

b 59

c 47

ACTIVITY

4 How much money altogether?

cents

 • *AUSTRALIAN SIGNPOST MATHS 1* • ISBN 978 1 4886 2178 9

17C Addition to 20

INVESTIGATION

1 Write number sentences about the picture.

☐ + ☐ = ☐   ☐ + ☐ = ☐

☐ + ☐ = ☐   ☐ + ☐ = ☐

☐ + ☐ = ☐   ☐ + ☐ = ☐

☐ + ☐ = ☐   ☐ + ☐ = ☐

17D Recognising 3D Objects

1 Match each picture with a 3D object in the centre.

2 **a** Name two shapes that are the faces of each 3D object.

A

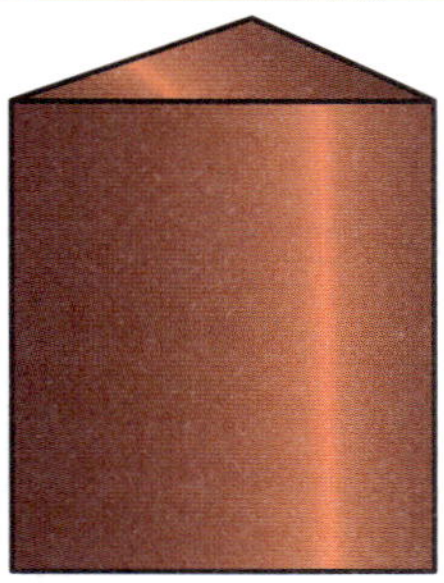

B

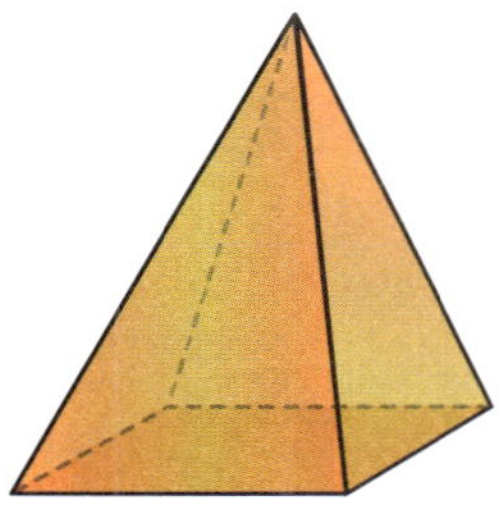

C

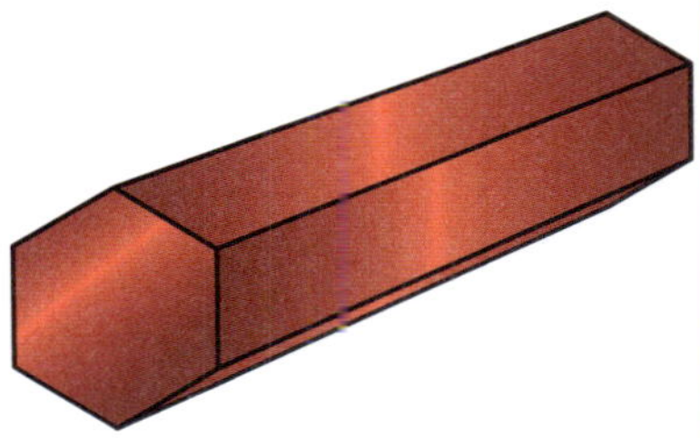

b How many corners does each object have?

A

B

C

 • *AUSTRALIAN SIGNPOST MATHS 1* • ISBN 978 1 4886 2178 9

18A Numbers to 100

1 Write the next two numbers.

a 37, ☐, ☐

b 86, ☐, ☐

c 40, ☐, ☐

d 64, ☐, ☐

e 79, ☐, ☐

2 Colour your answers from Questions 1 and 3 on the chart.

1	2	3	4	5	6	7	8	9	10
11	12	13	14	15	16	17	18	19	20
21	22	23	24	25	26	27	28	29	30
31	32	33	34	35	36	37	38	39	40
41	42	43	44	45	46	47	48	49	50
51	52	53	54	55	56	57	58	59	60
61	62	63	64	65	66	67	68	69	70
71	72	73	74	75	76	77	78	79	80
81	82	83	84	85	86	87	88	89	90
91	92	93	94	95	96	97	98	99	100

3 Write the missing number.

a 66, ☐, 68

b 34, ☐, 36

c 79, ☐, 81

d 92, ☐, 94

e 55, ☐, 57

f 88, ☐, 90

4 Count by tens to join the dots.

a

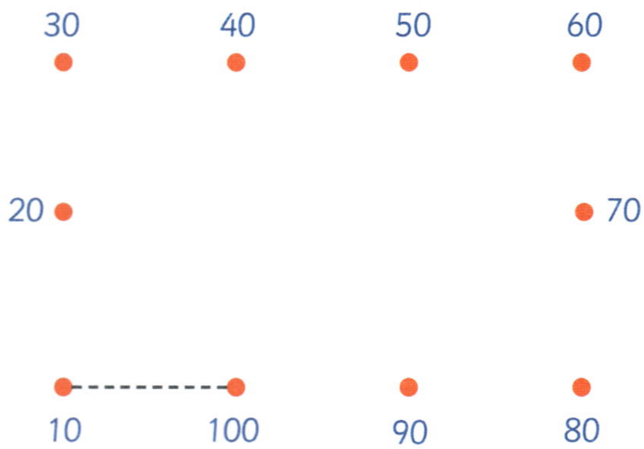

b

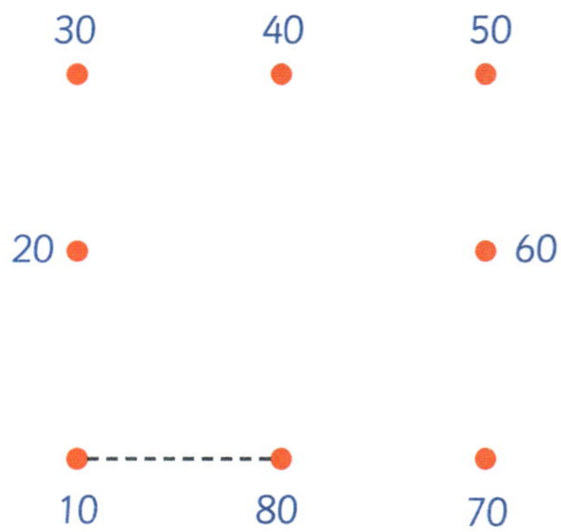

 • *AUSTRALIAN SIGNPOST MATHS 1* • ISBN 978 1 4886 2178 9

18B Partitioning

1 Colour the blocks to show two different ways to make 7.

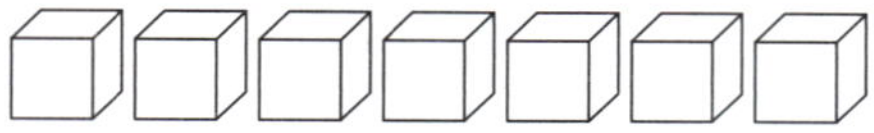

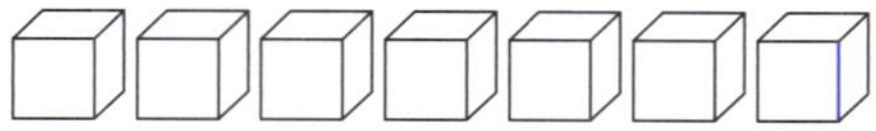

☐ and ☐ makes 7.   ☐ and ☐ makes 7.

2 Colour the circles to show two different ways to make 8.

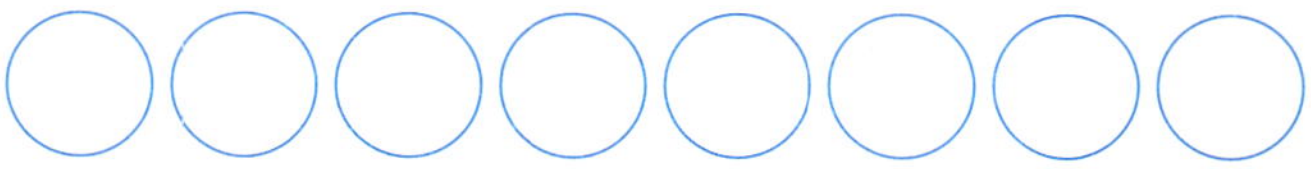

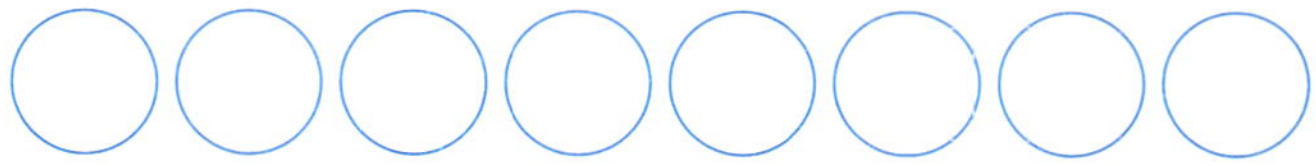

☐ + ☐ = 8   ☐ + ☐ = 8

3 Draw dominoes to show two different ways to make 9.

☐ + ☐ = 9   ☐ + ☐ = 9

4 Take 12 counters. Make two groups. Write a number sentence to show this. Make other number sentences for 12.

☐ + ☐ = ☐   ☐ + ☐ = ☐

☐ + ☐ = ☐   ☐ + ☐ = ☐

 • *AUSTRALIAN SIGNPOST MATHS 1* • ISBN 978 1 4886 2178 9

18C Partitioning

CONCEPT

4 + 2 = 6

We can find a different sum by moving the bar.

1 7

☐ + 6 = 7

2 + ☐ = 7

3 + 4 = ☐

4 + ☐ = 7

☐ + 2 = 7

6 + ☐ = 7

7

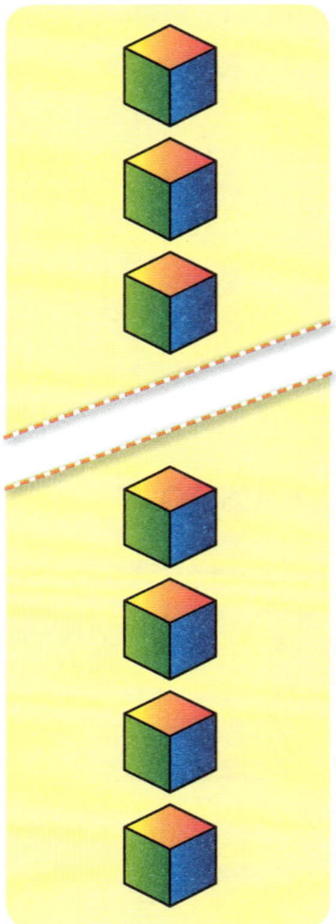

3 + 4

2 6

1 + 5 = ☐

2 + ☐ = 6

☐ + 3 = 6

4 + ☐ = 6

5 + 1 = ☐

3 Using counters, make up your own partitioning patterns for 10, 9, 8 and 5.

INVESTIGATION

You should know these.

3

1 + 2 = 3 →

2 + 1 = 3 →

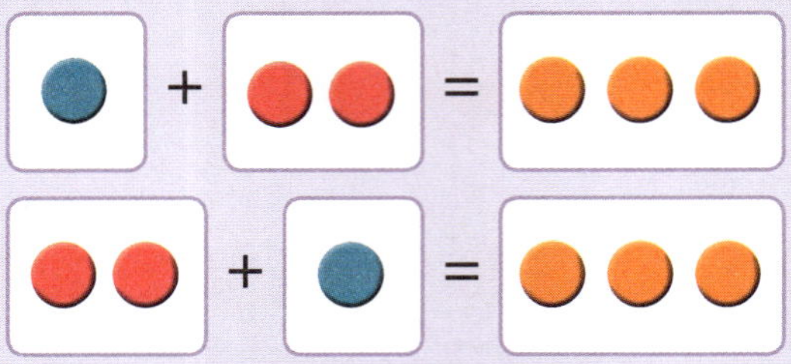

18D Half Past

Half past 9 is the same as 9:30.

1 Write the times shown.

a half past ☐

b half past ☐

c half past ☐

d half past ☐

e half past ☐

f half past ☐

2 Show these times.

a half past 3

b half past 10

c half past 1

19A Subtraction to 20

1. Count back to complete each number sentence.

a

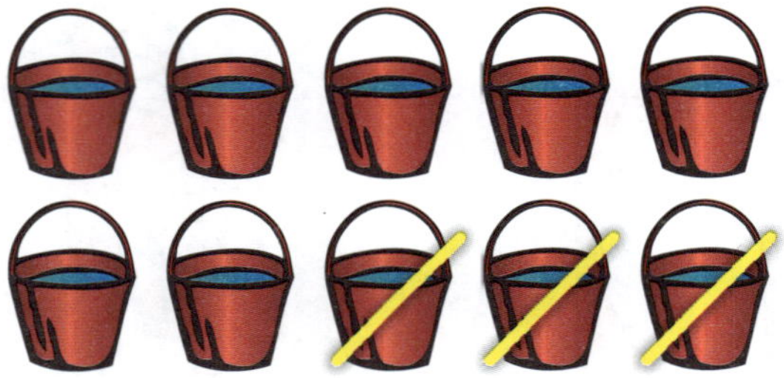

10 – 3 = ☐

b

11 – 6 = ☐

c

16 – 4 = ☐

d

15 – 9 = ☐

e

19 – 8 = ☐

f

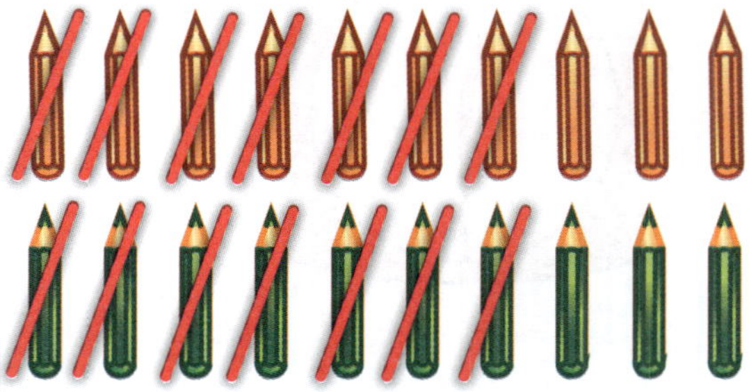

20 – 14 = ☐

2. **a** Count back 2.

11, ☐, ☐

14, ☐, ☐

b Count back 4.

17, ☐, ☐, ☐, ☐

20, ☐, ☐, ☐, ☐

19B Subtraction by Counting On

CONCEPT

How many more make 8?

- I count on from 6 to make 8.
- I need 2 more to make 8.

$6 + \square = 8$

$8 - 6 = \square$

1 Draw the missing objects and complete.

a

How many more to make 8?

$6 + \square = 8$ $8 - 6 = \square$

b

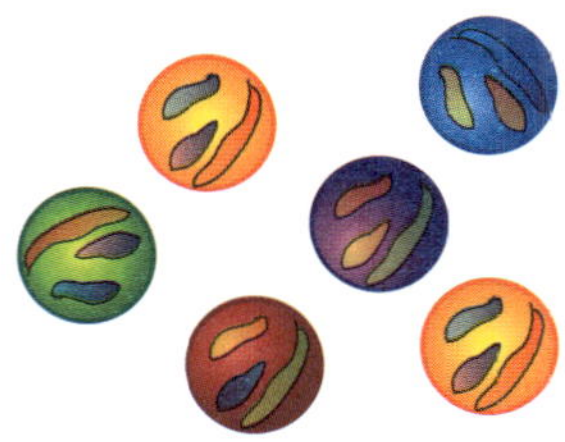

How many more to make 7?

$6 + \square = 7$ $7 - 6 = \square$

c

How many more to make 9?

$6 + \square = 9$ $9 - 6 = \square$

d

How many more to make 10?

$7 + \square = 10$ $10 - 7 = \square$

e

How many more to make 8?

$5 + \square = 8$ $8 - 5 = \square$

f

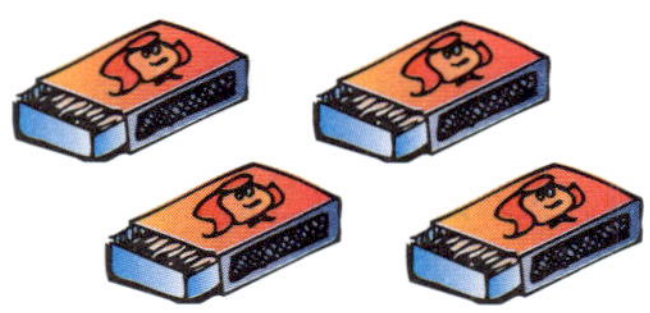

How many more to make 7?

$4 + \square = 7$ $7 - 4 = \square$

 • *AUSTRALIAN SIGNPOST MATHS 1* • ISBN 978 1 4886 2178 9

19C One Half

CONCEPT

One half is one of two equal parts.

1. Draw a line to cut each object into halves. Colour one half.

2. If half of the group is coloured, tick the box.

a  ☐

b  ☐

c 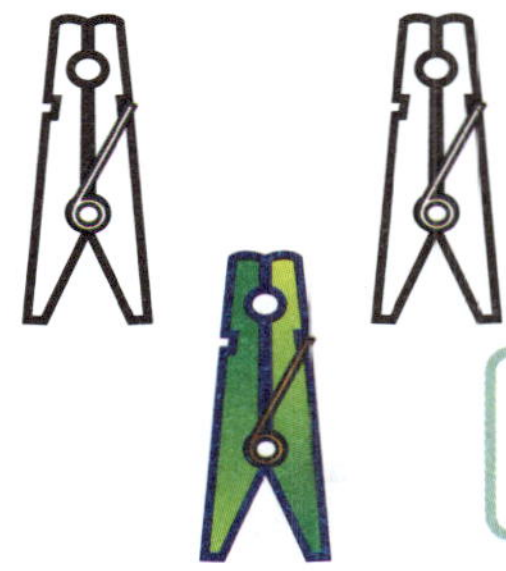 ☐

3. Colour half of each group.

a 

b 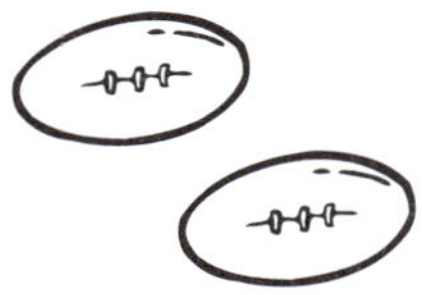

c 

d 

FUN SPOT

- Count 20 counters.
- Share the counters into two equal groups.
- Count each group.
- Find halves of other groups.

19D Half of a Group

$\frac{1}{2}$

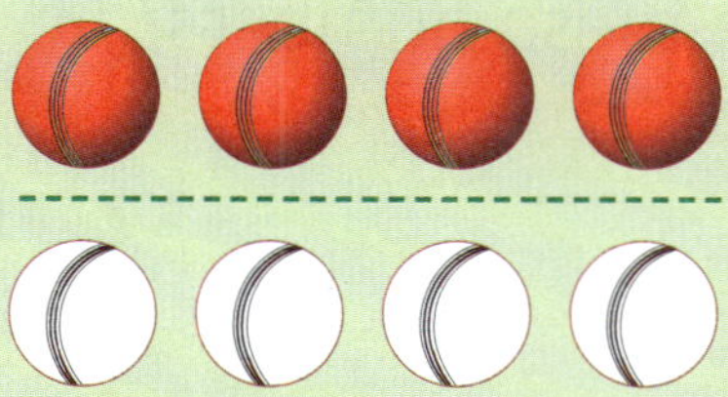

halves

One half is 1 of 2 equal parts.

1 Circle the groups that are divided into halves.

Colour one half of each group.

a
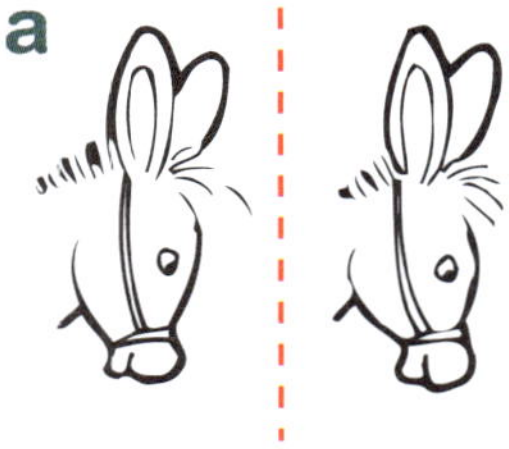

b
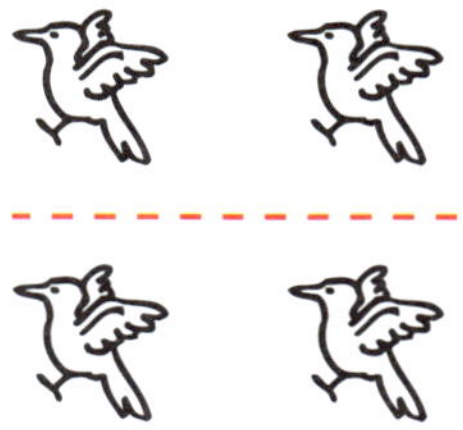

c
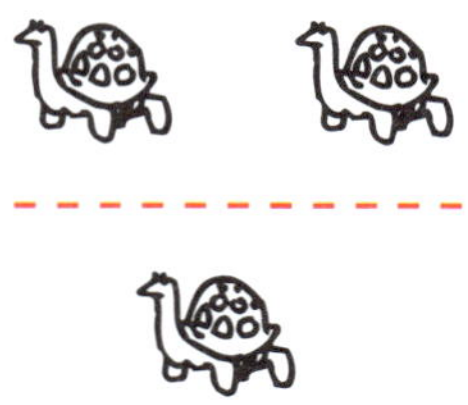

d
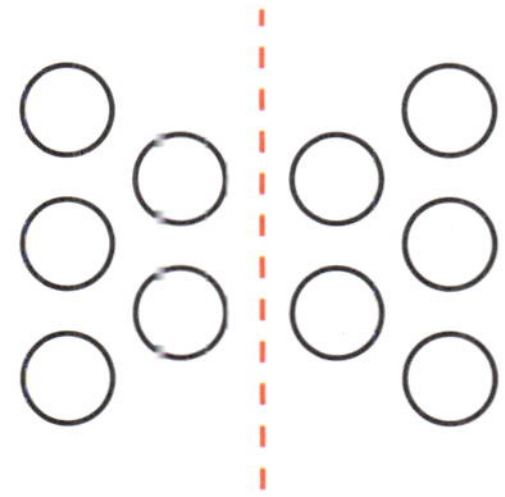

e
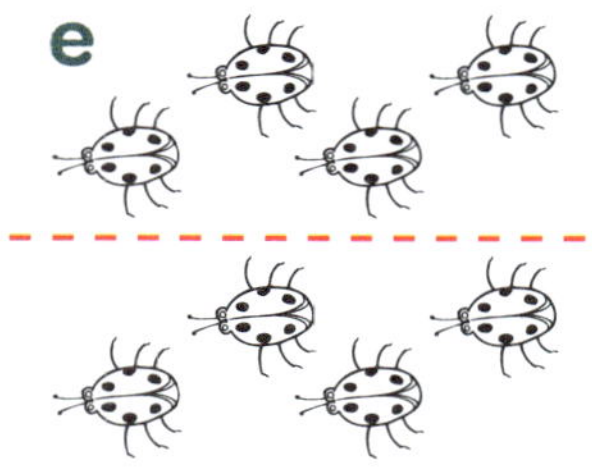

f
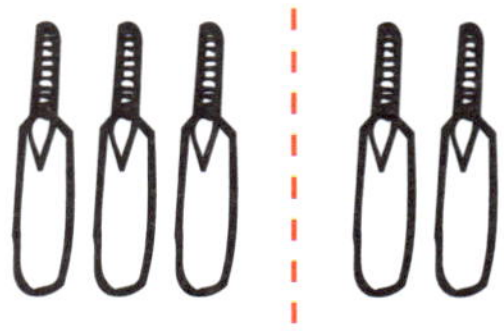

g

h
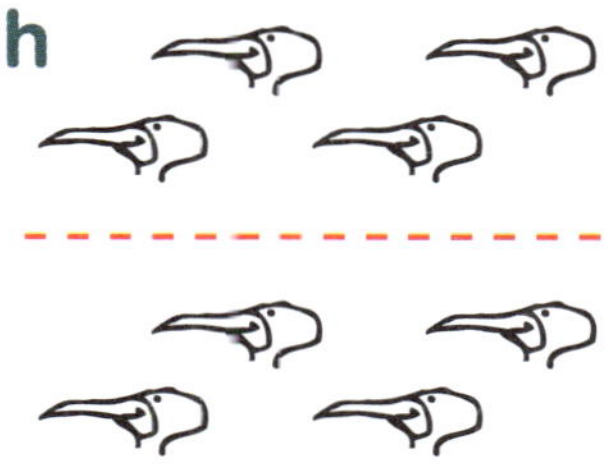

2 Circle half of each group.

a
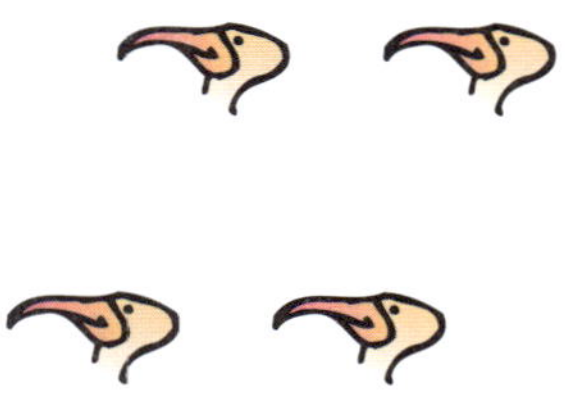

b

c
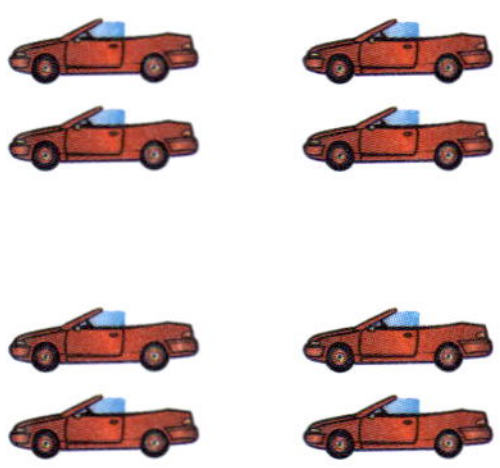

d

e

f

g
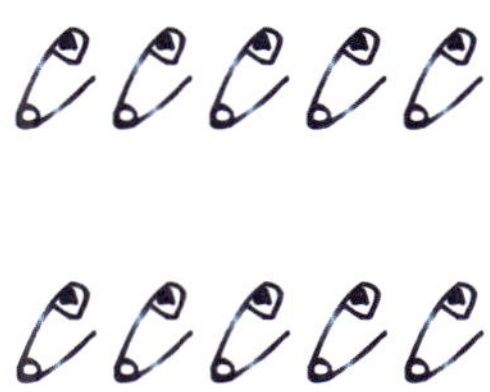

h
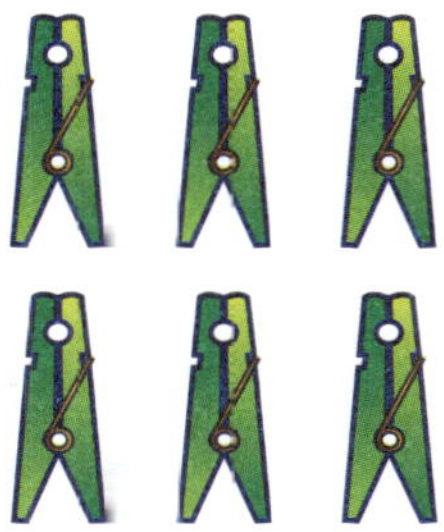

20A Numbers to 100

1 Use numerals to write these numbers.

a ten ☐ **b** forty ☐

c fifty-two ☐ **d** nineteen ☐

e one hundred ☐ **f** seventy-three ☐

2 Write words for these numbers.

a 21 ____________________

b 100 ____________________

c 54 ____________________

Don't forget to use a hyphen for twenty-one.

3 Match the numerals and words.

a

fifty-one	83
100	sixty
60	51
eighty-three	one hundred

b

37	40
forty	seventy-five
ninety-nine	thirty-seven
75	99

Draw lines to match.

 • *AUSTRALIAN SIGNPOST MATHS 1* • ISBN 978 1 4886 2178 9

20B Counting by 2s, 5s and 10s

1	2	3	4	5	6	7	8	9	10
11	12	13	14	15	16	17	18	19	20
21	22	23	24	25	26	27	28	29	30
31	32	33	34	35	36	37	38	39	40
41	42	43	44	45	46	47	48	49	50

1. Count by twos to finish the pattern.

   2, 4, 6, ☐, ☐, ☐, ☐, ☐, ☐, ☐

1	2	3	4	5	6	7	8	9	10
11	12	13	14	15	16	17	18	19	20
21	22	23	24	25	26	27	28	29	30
31	32	33	34	35	36	37	38	39	40
41	42	43	44	45	46	47	48	49	50

2. Count by fives to finish the pattern.

   5, 10, 15, ☐, ☐, ☐, ☐, ☐, ☐

3. Count by tens to finish the pattern.

   10, 20, 30, ☐, ☐, ☐, ☐, ☐, ☐

4. Look at the beads around the page. Colour every 5th bead.

Start

 AUSTRALIAN SIGNPOST MATHS 1 • ISBN 978 1 4886 2178 9

20C Looking at Patterns

CONCEPT

We can make patterns by adding or subtracting.

- Start with 0 and **add 5**: 0, 5, 10, 15, 20, …
- Start with 90 and **subtract 10**: 90, 80, 70, …

1 Complete the number patterns.

a 12, 14, 16, ___, ___, ___, ___, ___

b 10, 20, 30, ___, ___, ___, ___, ___

c 25, 24, 23, ___, ___, ___, ___, ___

2 We can write the rule for Question 1**a** above as **add 2**.

The rule for 1**b** is ______________.

The rule for 1**c** is ______________.

3 Write the rule for the pattern on the number line.

a

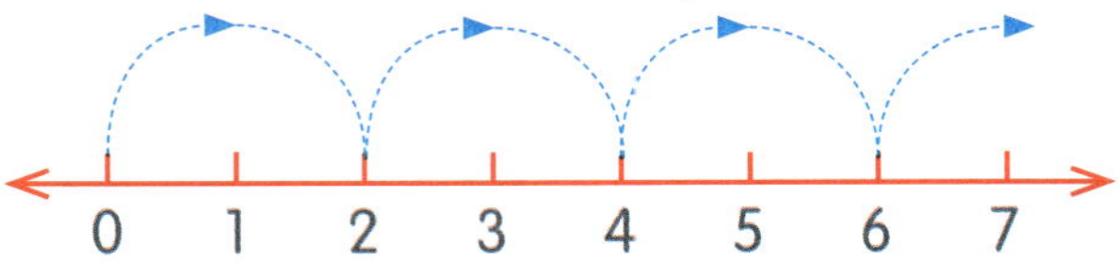

____

b

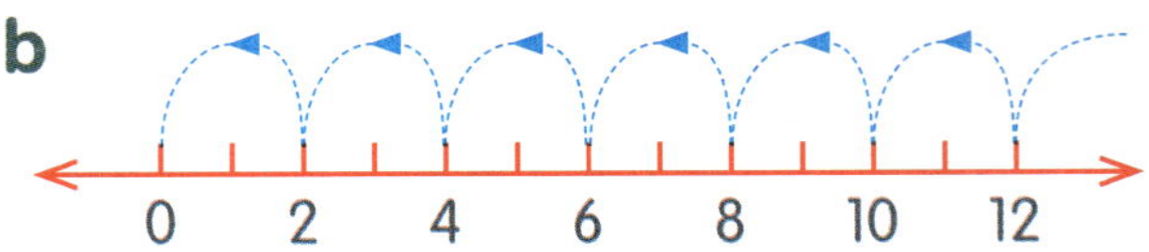

____

INVESTIGATION

Use counters to make a pattern.

What rule did you use? ______________

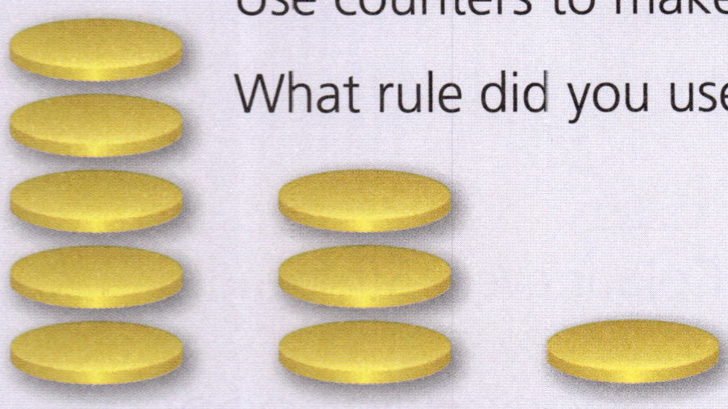

Make other patterns using blocks.

20D Half Past

half past 2

On digital clocks, **half past** times always have **30** as the last two digits.

half past 2

CONCEPT

1 Read the time. Complete the label.

a

half past ☐

b

half past ☐

c

half past ☐

d

half past ☐

e

half past ☐

f

half past ☐

2 Write the time.

a

half past ☐

b

half past ☐

c

half past ☐

3 Write the digital time.

a half past 3 ☐

b half past 5 ☐

 • *AUSTRALIAN SIGNPOST MATHS 1* • ISBN 978 1 4886 2178 9

21A Number Relationships

CONCEPT

= can mean **is the same as.** **2 + 2 = 1 + 3 = 4**

1. Use partitioning with blocks to show the relationships.

a 1 + 4 = ☐ 2 + 3 = ☐

b 5 + 2 = ☐ 3 + 4 = ☐

c 8 + 5 = ☐ 4 + 9 = ☐

d 5 + 4 = ☐ 3 + 6 = ☐

e 1 + 4 = 2 + 3 = ☐

f 5 + 2 = 3 + 4 = ☐

2.

a 3 + 1 = ☐ 1 + 3 = ☐

b 6 + 2 = ☐ 2 + 6 = ☐

c 4 + 5 = ☐ 5 + 4 = ☐

d 7 + 3 = ☐ 3 + 7 = ☐

e 3 + 1 = 1 + 3 = ☐

f 6 + 2 = 2 + 6 = ☐

g 4 + 5 = 5 + 4 = ☐

h 9 + 3 = 3 + 9 = ☐

ICT

Use a calculator to find additions that have the same answer.

Share these with your partner.

21B Adding Equal Groups

CONCEPT

5 + 5 + 5

3 rows of 5 = 15

1 **a**

☐ + ☐

☐ rows of ☐ = ☐

b

☐ + ☐ + ☐

☐ rows of ☐ = ☐

c

☐ + ☐ + ☐ + ☐

☐ groups of ☐ = ☐

d

☐ + ☐ + ☐

☐ groups of ☐ = ☐

e 4 groups of 2 trees

☐ + ☐ + ☐ + ☐

☐ groups of ☐ = ☐

f 3 rows of 4 worms

☐ + ☐ + ☐

☐ rows of ☐ = ☐

21C Graphing Data Using Blocks

1 After each game, we put a block next to the winner's name.

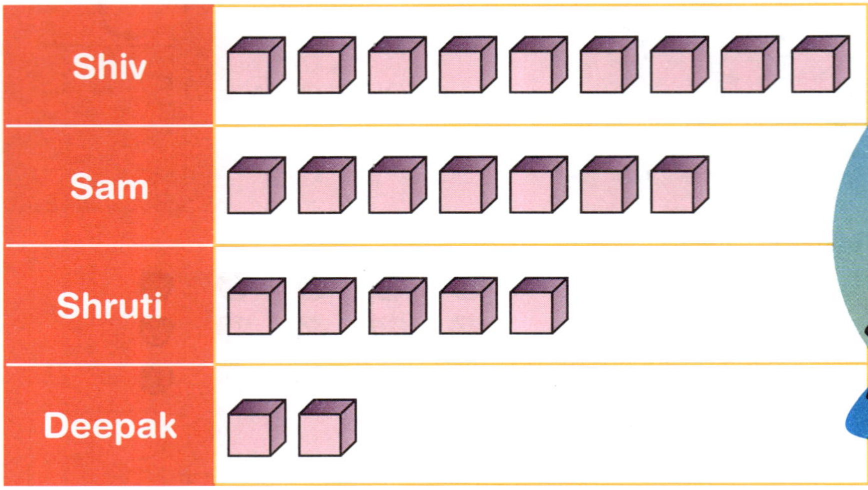

a Who won the most games?

b Who won the least games?

c How many games did Sam win?

d How many games have been played?

Ask 10 students to choose their favourite colour out of red, blue and green.

Put blocks or counters on the graph below to show their answers.

Red							
Blue							
Green							

 • *AUSTRALIAN SIGNPOST MATHS 1* • ISBN 978 1 4886 2178 9

21D Graphs

Everybody brought their pets to school.

1. Look at the picture and complete this table.

Pets Brought to School			
Birds	Cats	Dogs	Rabbits

2. Draw a graph of the pets.

Pets Brought to School					
Birds					
Cats					
Dogs					
Rabbits					

3. Write about the graph.

____

____

INVESTIGATION

What pets do people in your class have? Find out and fill in this graph.

Birds									
Cats									
Dogs									
Fish									
Rabbits									
Other									

 • *AUSTRALIAN SIGNPOST MATHS 1* • ISBN 978 1 4886 2178 9

22A Partitioning

CONCEPT

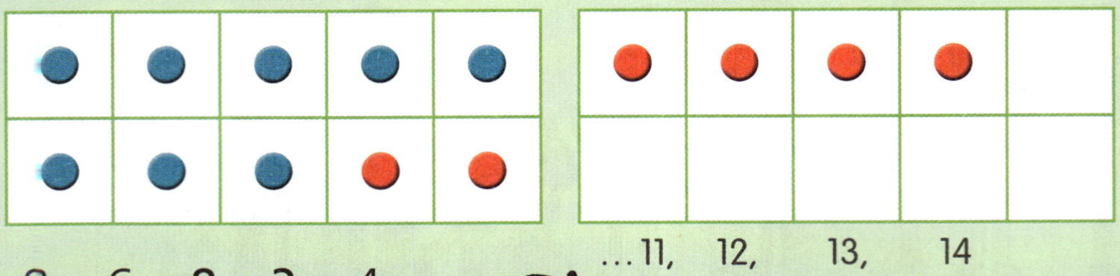

$8 + 6 = \mathbf{8 + 2 + 4}$

$= 10 + 4$

$= 14$

1 To find the answer, draw more dots in the ten frames.

a

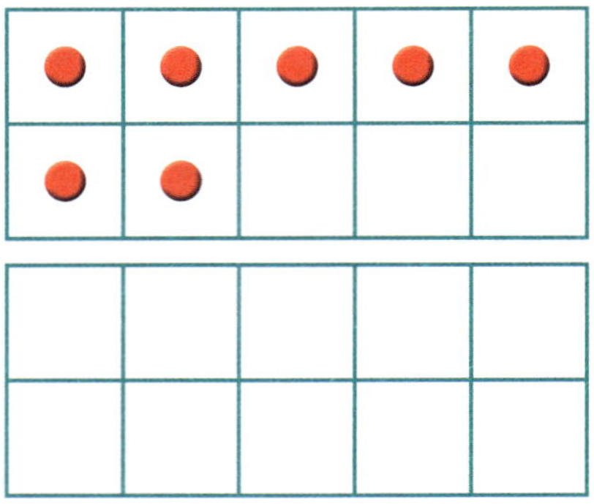

7 + 5 = 7 + ☐ + ☐

= ☐ + ☐ = ☐

b

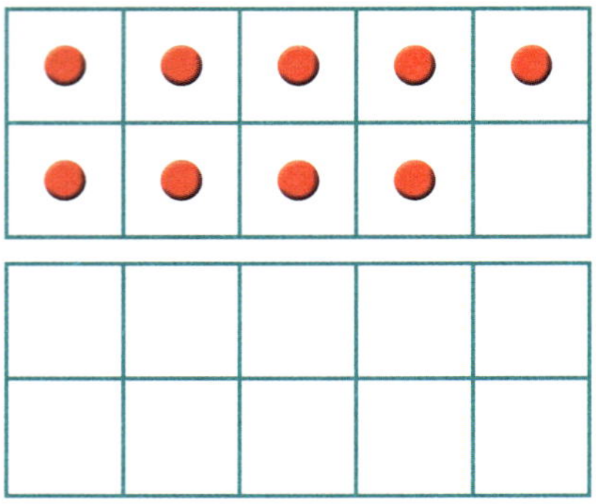

9 + 8 = 9 + ☐ + ☐

= ☐ + ☐ = ☐

2 You can use ten frames to answer these questions.

a 7 + 6 = ☐

b 8 + 5 = ☐

c 9 + 4 = ☐

d 8 + 7 = ☐

INVESTIGATION

Use place-value blocks to make your own patterns. Record each pattern and explain your patterns to a friend.

22B Partitioning

1 To find the answer, draw more dots in the ten frames.

a

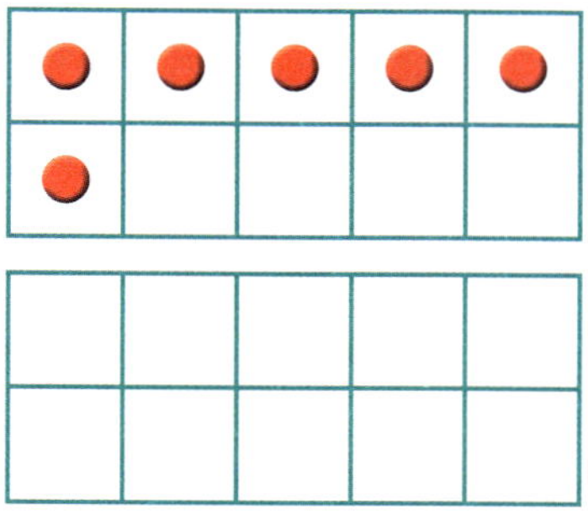

$6 + 9 = 6 + \square + \square$

$= \square + \square = \square$

b

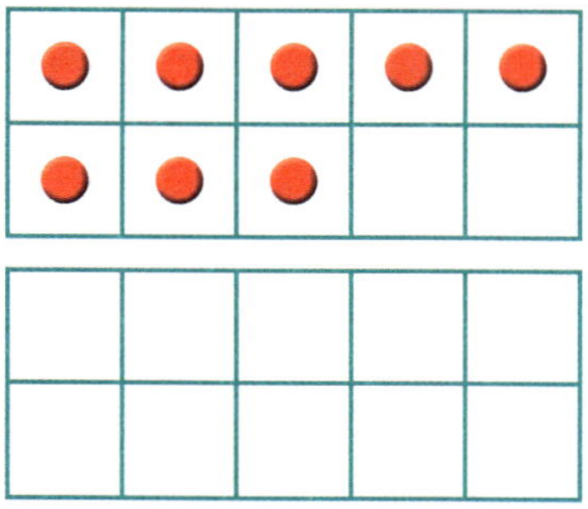

$8 + 8 = 8 + \square + \square$

$= \square + \square = \square$

c

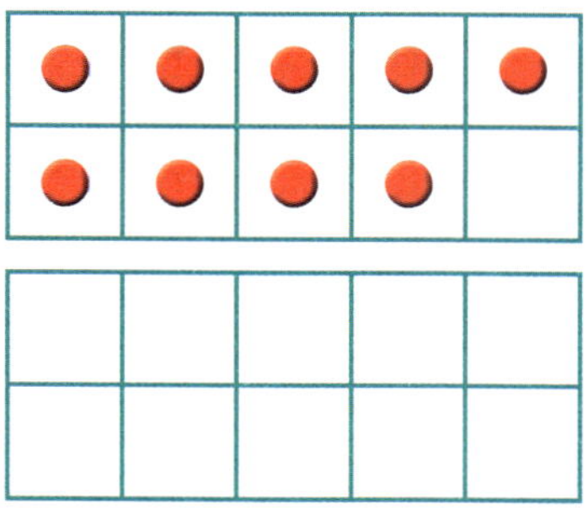

$9 + 7 = 9 + \square + \square$

$= \square + \square = \square$

d

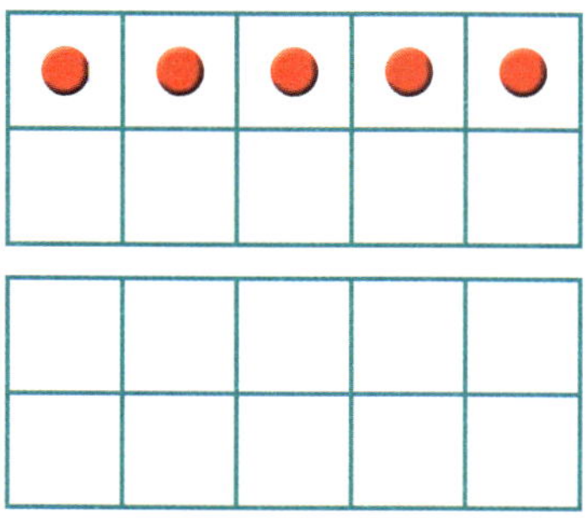

$5 + 6 = 5 + \square + \square$

$= \square + \square = \square$

Make groups of ten, then count on.

INVESTIGATION

Use ten frames to make up some additions of your own.

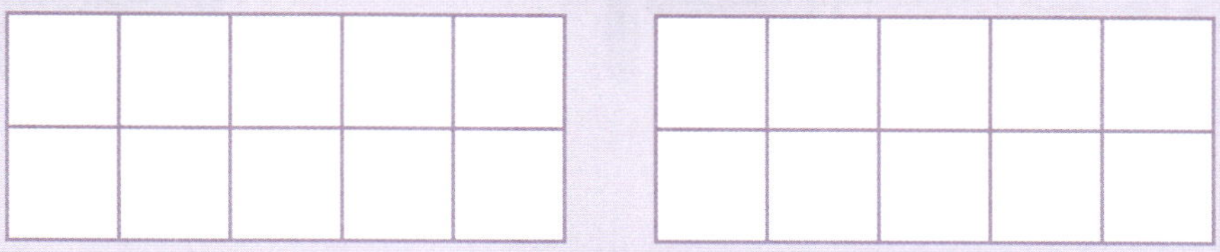

 • *AUSTRALIAN SIGNPOST MATHS 1* • ISBN 978 1 4886 2178 9

22C Informal Units of Length

ACTIVITY

1 Use different units to measure length.

Length	Tally	Measurement Unit	
Length of Pencil	||||	4	toothpicks
Height of Desk			hand spans
Width of Book			paperclips
Width of Door			straws
Length of Room			books

2 Estimate and then measure these lengths using craft sticks.

a

Estimate | Measure

b

Estimate | Measure

c

Estimate | Measure

d

Estimate

Measure

e

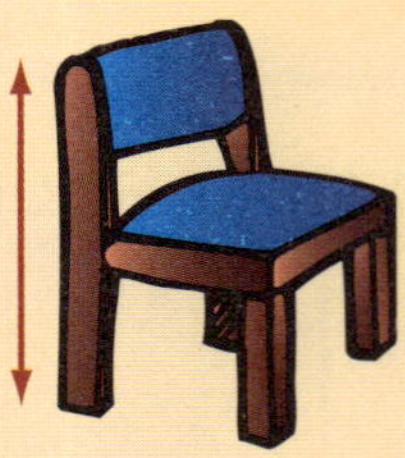

Estimate

Measure

f

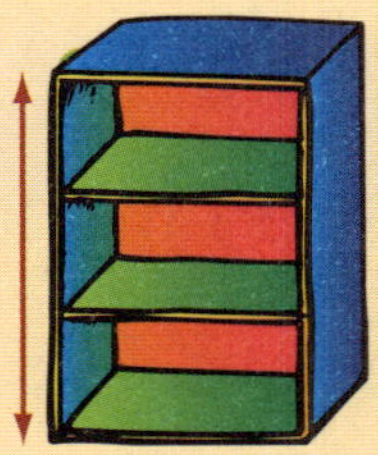

Estimate

Measure

 ISBN 978 1 4886 2178 9

22D Informal Units of Length

ACTIVITY

Use parts of your body to measure objects.
Compare your results with your partner's.

Length	Unit	Estimate	Measurement	Partner's Measurement
Width of Book	hand span			
Width of Book	finger			
Width of Window	hand span			

Using a Standard Informal Unit

- Choose a standard length unit (such as an eraser).
- Mark this unit's lengths (1 eraser, 2 erasers, 3 erasers, ...) along a ribbon or a piece of string.
- Use your ribbon or string to measure lengths.
- Record your results.

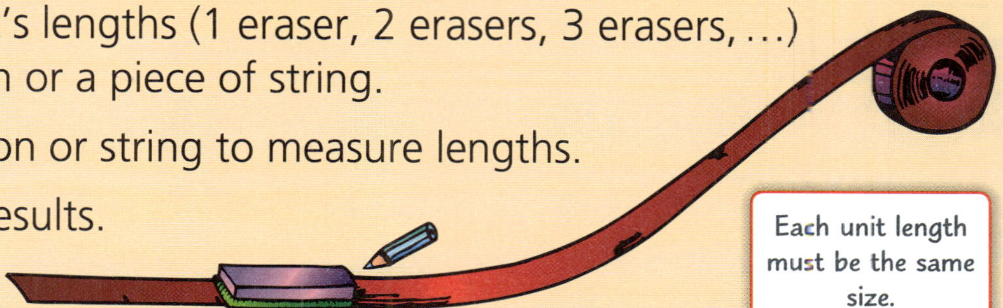

Length	Estimate	Measurement
Length of Table	units	units
Width of Door	units	units
Width of Cupboard	units	units

Would the cupboard fit through the door?

 • *AUSTRALIAN SIGNPOST MATHS 1* • ISBN 978 1 4886 2178 9

23A Place Value

CONCEPT

34

3 Tens 4 Ones

thirty-four

1 Write the number shown by each numeral expander.

a 6 Tens 5 Ones

b 7 Tens 0 Ones

c 5 Tens 7 Ones

d 9 Tens 2 Ones

e 4 Tens 6 Ones

f 8 Tens 3 Ones

2 Complete:

a 78 = ☐ tens ☐ ones

b 80 = ☐ tens ☐ ones

c 54 = ☐ tens ☐ ones

d 61 = ☐ tens ☐ one

3 Circle the smaller number.

a 91 or 89

b 62 or 67

c 42 or 58

FUN SPOT

Work with a partner and take turns.

One player picks up two number cards and puts them next to each other. The other player says this number.

 • *AUSTRALIAN SIGNPOST MATHS 1* • ISBN 978 1 4886 2178 9

Place Value

1 Complete each numeral expander and write the number.

a
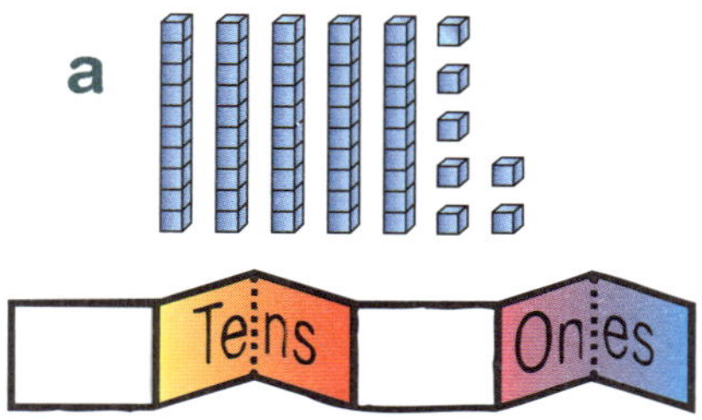

b
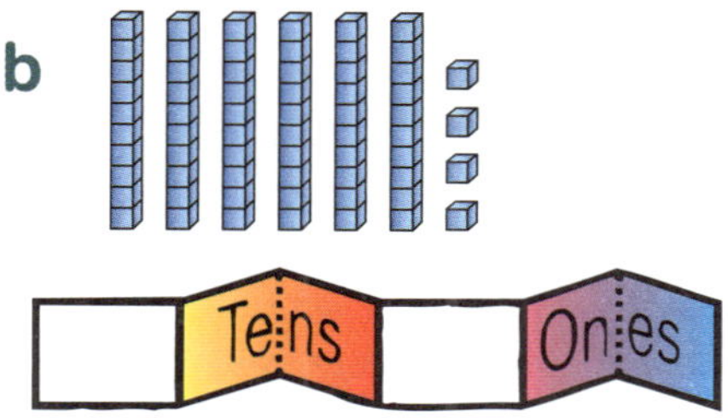

c
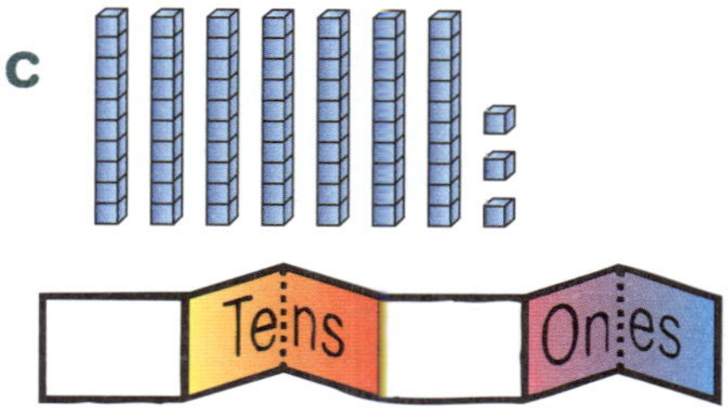

d
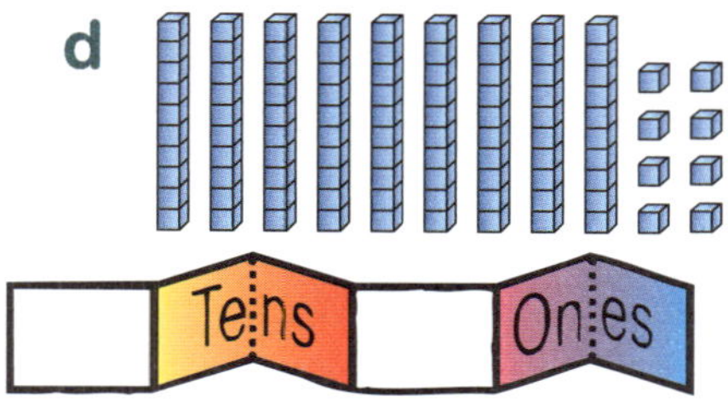

e
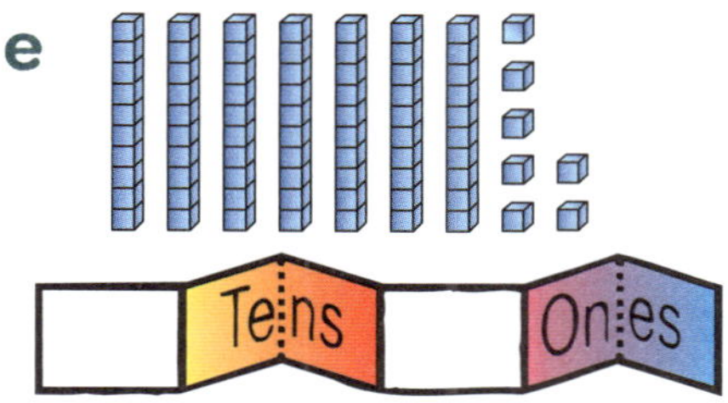

f
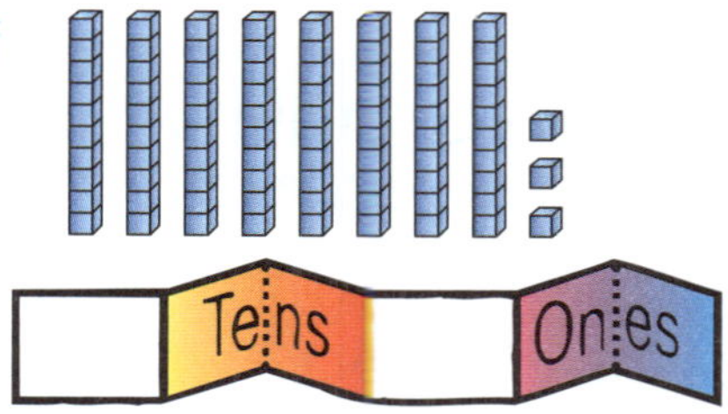

2 Write the numeral.

a fifty-one

b thirty-eight

c eighty-two

d forty-seven

e ninety-four

f fourteen

3 Write the numbers before and after.

a

Before		After
	78	
	35	

b

Before		After
	80	
	56	

23C The Rhombus

CONCEPT

This is a **rhombus**.

It is also called a diamond.

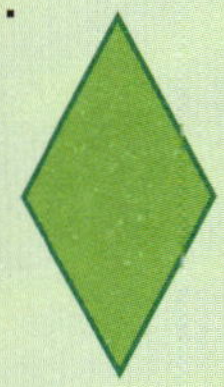

1 Name the shapes used.

Shape Words

Circle Triangle Square Rectangle Hexagon

2 Trace around the shape. Write its name.

3 Colour the rhombuses red.

4 Draw a rhombus.

5 A rhombus has ☐ sides and ☐ corners.

6 Circle the kites on this page.

 • *AUSTRALIAN SIGNPOST MATHS 1* • ISBN 978 1 4886 2178 9

23D 2D Shapes

Shape Words

Circle Hexagon Oval Triangle Square Rectangle Rhombus

1 Join the dots. Write the shape name and the number of sides and corners.

a

Sides | Corners

Shape Name:

b

Sides | Corners

Shape Name:

c

Sides | Corners

Shape Name:

d

Sides | Corners

Shape Name:

INVESTIGATION

Draw as many shapes as you can.

24A Numbers to 100

1. What number is shown by each numeral expander?

a
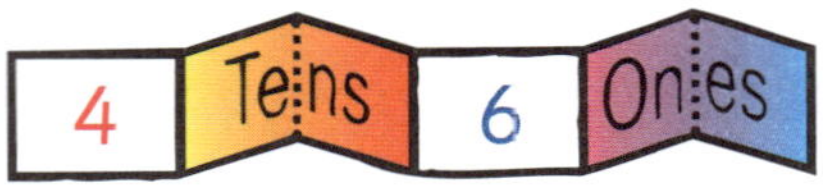

b

c

d

2. Expand the numbers.

a 90 = ☐ tens ☐ ones

b 73 = ☐ tens ☐ ones

c 62 = ☐ tens ☐ ones

3. Write the numerals for these numbers.

a sixty ☐

b eighty-three ☐

c thirty-seven ☐

d one hundred ☐

FUN SPOT

Every 10th bead around this page is shaded.

4. Count the beads. Colour the beads at these numbers:

a 16 b 25 c 48 d 67 e 88 f 100

Start

24B Place Value

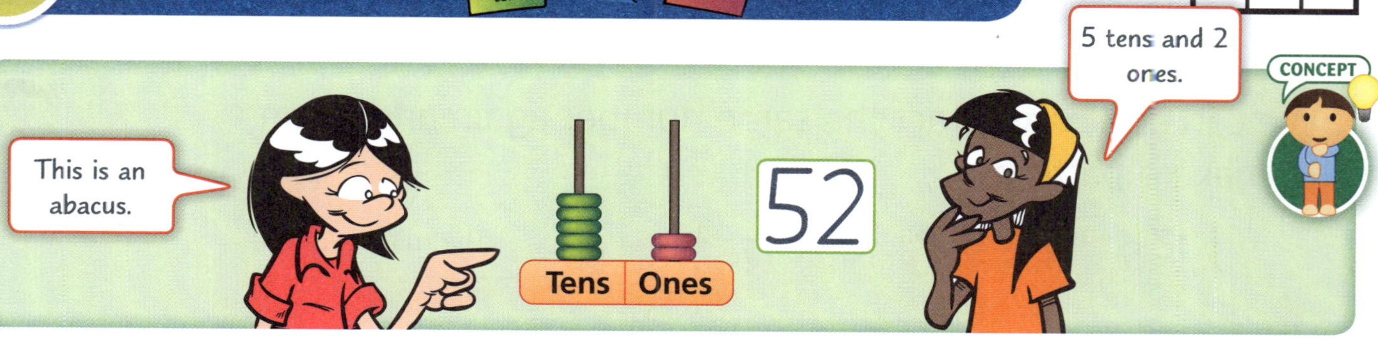

1 What number is shown on each abacus?

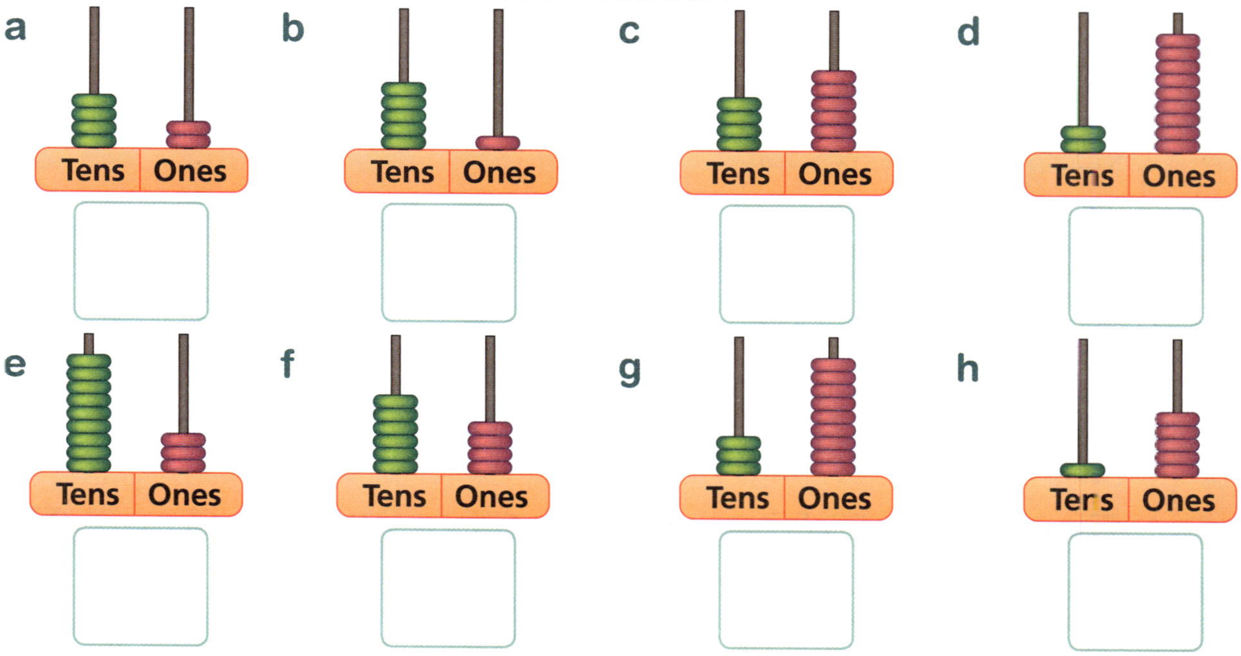

2 Complete each numeral expander and write the number.

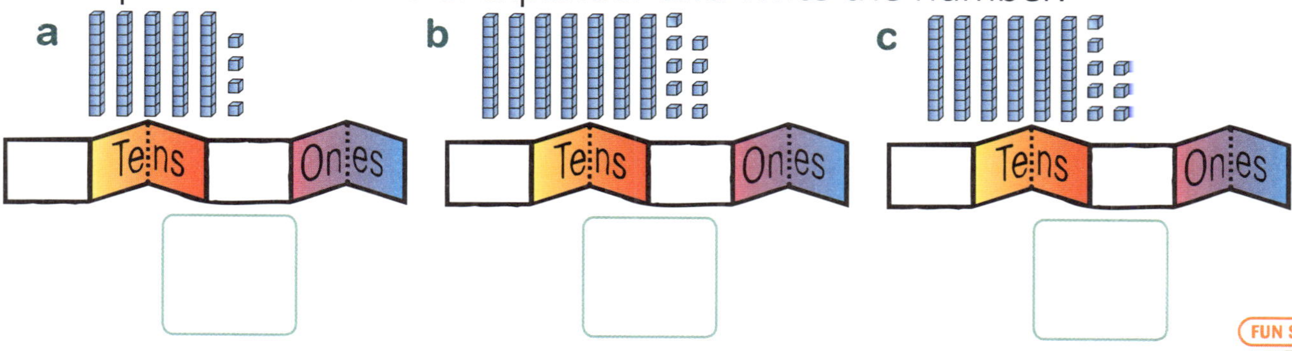

Take turns to roll two dice. Take the same number of cubes as your dice roll.

Make towers of 10. The first person to reach 50 wins!

 • *AUSTRALIAN SIGNPOST MATHS 1* • ISBN 978 1 4886 2178 9

24C Patterns

We can add or subtract the same number again and again to make a pattern.

2, $^{+3}$ 5, $^{+3}$ 8, $^{+3}$ 11, $^{+3}$ 14, $^{+3}$ 17, $^{+3}$ …

The rule is +3.

1 Complete the number patterns.

a 19 (+1) 20 (+1) 21 (+1) ☐ (+1) ☐ ☐ ☐ ☐

b 8 (+2) 10 (+2) 12 (+2) 14 (+2) ☐ ☐ ☐ ☐

c 80 (-10) 70 (-10) 60 (-10) 50 (-10) ☐ ☐ ☐ ☐

d 4 (+3) 7 (+3) 10 (+3) 13 (+3) ☐ ☐ ☐ ☐

2 Write the next two numbers.

a 28, 29, 30, ☐, ☐

b 9, 8, 7, ☐, ☐

c 7, 9, 11, ☐, ☐

d 0, 3, 6, ☐, ☐

e 12, 10, 8, ☐, ☐

f 50, 49, 48, ☐, ☐

Write a number pattern of your own.

☐ ☐ ☐ ☐ ☐ ☐ ☐ ☐

 • *AUSTRALIAN SIGNPOST MATHS 1* • ISBN 978 1 4886 2178 9

24D Months of the Year

January, February, March, April, May, June, July, August, September, October, November, December

❶ Discuss the calendar below.

January	February	March	April
	School starts	Tom's birthday	Easter Holidays
May	**June**	**July**	**August**
	Ella's birthday		
September	**October**	**November**	**December**
		Speech Night	Christmas Holidays

a Which month is Tom's birthday? ☐

b How many months from Tom's birthday until Christmas? ☐

c How many months from January to Easter? ☐

d If it is August now, how many months ago was Ella's birthday? ☐

e If it is August now, how many months ago was January? ☐

f Write your own birthday and other events on the calendar.

© PEARSON AUSTRALIA 2018 • *AUSTRALIAN SIGNPOST MATHS 1* • ISBN 978 1 4886 2178 9

Counting On

6 + 3 = 9

You can start with 6 and then use your fingers to count on 3 more.

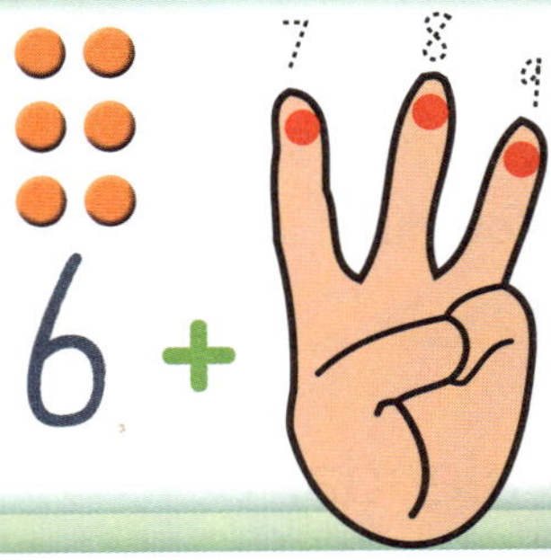

1 Use fingers to count on.

a 5 +

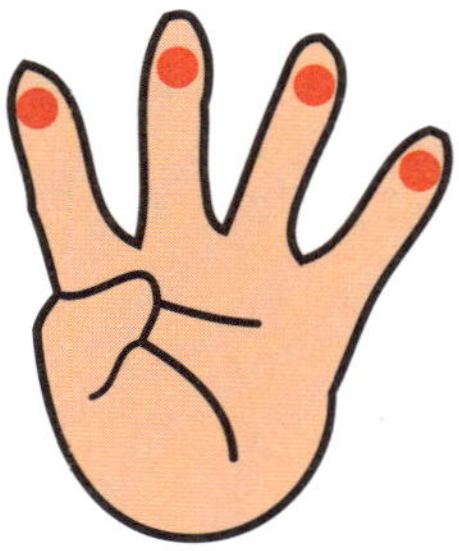

5 + 4 = ☐

b 8 +

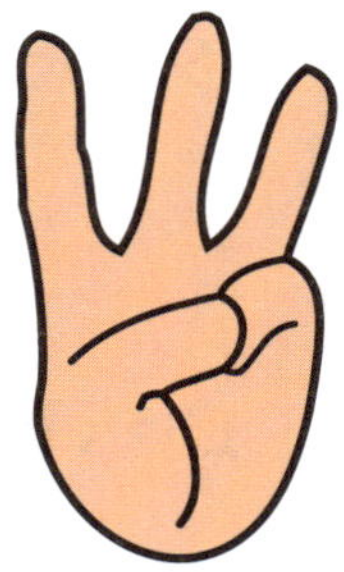

8 + 3 = ☐

c 9 +

9 + 2 = ☐

d 13 +

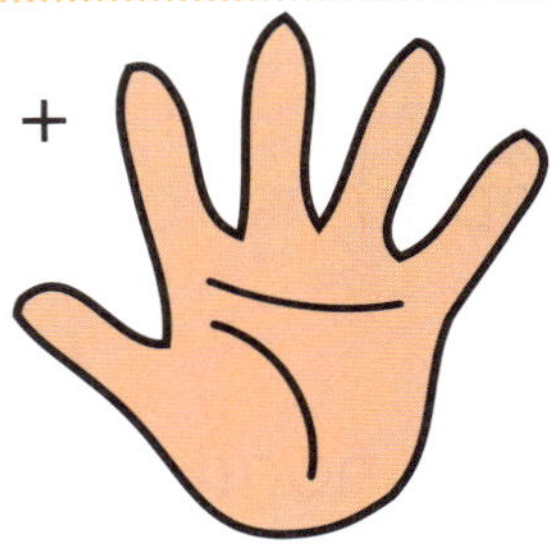

13 + 5 = ☐

2 Use your own fingers to count on.

a 7 + 4 = ☐

b 14 + 4 = ☐

c 9 + 3 = ☐

d 17 + 2 = ☐

Make up more questions of your own.

 ISBN 978 1 4886 2178 9

25B Addition by Counting On

CONCEPT

11 + 3 = 14

11 count on 3 → 12, 13, 14.

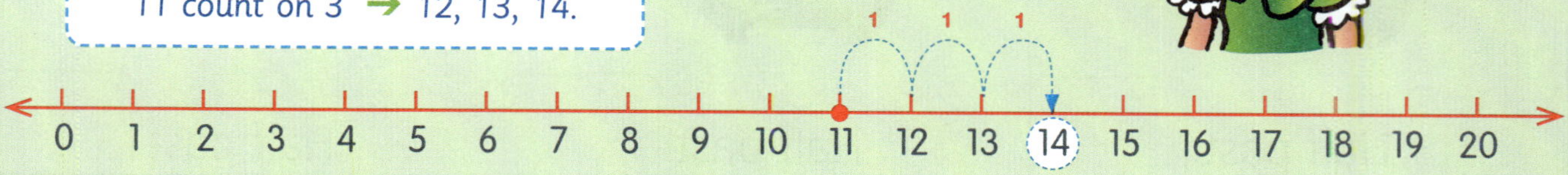

1 Use the dots to count on.

a 8 + 1 = ☐　**b** 9 + 2 = ☐　**c** 17 + 1 = ☐

d 12 + 2 = ☐　**e** 18 + 1 = ☐　**f** 15 + 2 = ☐

g 9 + 3 = ☐　**h** 8 + 2 = ☐　**i** 11 + 4 = ☐

j 14 + 2 = ☐　**k** 17 + 3 = ☐　**l** 19 + 1 = ☐

2 Use the number line to find the answers.

a 8 + 4 = ☐　**b** 9 + 5 = ☐

c 16 + 3 = ☐　**d** 7 + 6 = ☐

e 12 + 4 = ☐　**f** 17 + 4 = ☐

25C Digital and Analogue Time

1 Read and write each time.

a

Half past ☐

b

Half past ☐

c

Half past ☐

2 Write the digital time.

a Half past 8 ☐

b Half past 6 ☐

c Half past 12 ☐

3 Read and write each time.

a

Half past ☐

b

Half past ☐

c

Half past ☐

4 Draw the time on each clock.

a Half past 2

b Half past 11

c Half past 5

 • *AUSTRALIAN SIGNPOST MATHS 1* • ISBN 978 1 4886 2178 9

25D Making a Graph

1 Matt tossed a coin ten times and made this graph.

Heads							
Tails							

a How many  ? ☐

b How many  ? ☐

c Which was tossed more often? ☐

INVESTIGATION

2 Toss a coin eight times. Colour heads or tails after each throw.

Heads							
Tails							

a How many  ? ☐

b How many  ? ☐

c Which was tossed more often? ☐

FUN SPOT

Make a graph of the colours you like and the colours you don't like.

Colours I like				
Colours I don't like				

26A Counting Back

14 – 4 = 10

Start at 14.
Jump back 4.

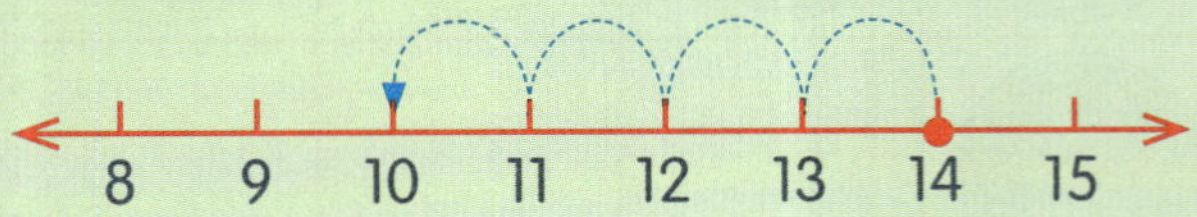

CONCEPT

1 Use these number lines to count back.

a

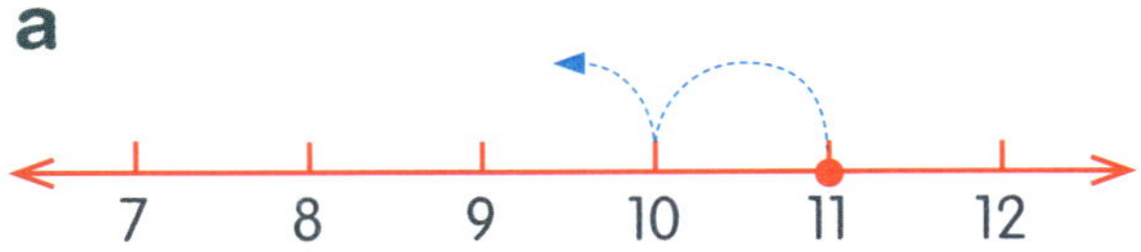

11 – 4 = ☐

b

15 – 5 = ☐

c

12 – 3 = ☐

d

16 – 6 = ☐

e

17 – 2 = ☐

f

19 – 4 = ☐

ACTIVITY

Use this number line to create your own subtraction sentences.

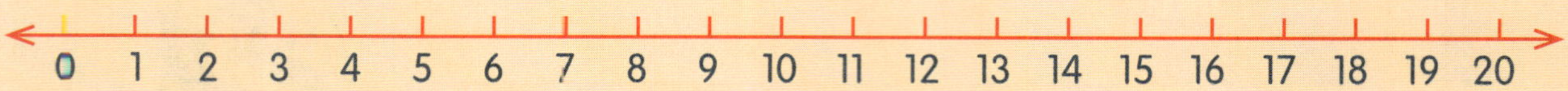

 ISBN 978 1 4886 2178 9

26B Subtraction

CONCEPT

Count back 6 from 11.

11 – 6 = 5

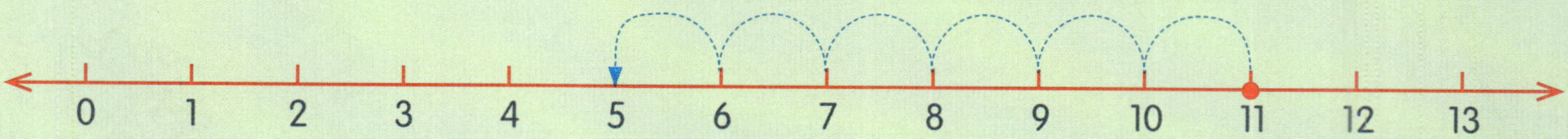

1 Use the number line to complete each number sentence.

a 6 – 3 = ☐ **b** 13 – 6 = ☐ **c** 9 – 5 = ☐

d 14 – 9 = ☐ **e** 16 – 8 = ☐ **f** 15 – 9 = ☐

g 17 – 0 = ☐ **h** 20 – 5 = ☐ **i** 19 – 7 = ☐

2 **a** Ruby had $15.
She spent $4.
How much does she have left?

☐ – ☐ = ☐

b Matt had 18 worms.
10 worms died.
How many worms are left?

☐ – ☐ = ☐

3 Colour the words that mean **–**.

take away | add | plus | difference between | less than | subtract | minus

 • *AUSTRALIAN SIGNPOST MATHS 1* • ISBN 978 1 4886 2178 9

26C Properties of Shapes

1 **a** Complete this table.

Shape	Name	Number of Corners	Number of Sides
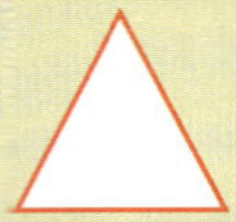			
		4	

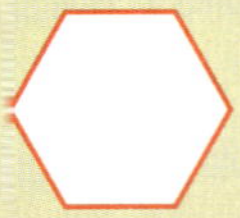			
	circle		
			1

Hexagons have 6 sides and 6 corners.

b Colour in the shapes that have all sides equal.

c Circle the shapes that have four sides.

d Write the names of the shapes that have:

3 corners ☐ 6 sides ☐

2 This is a 5-piece **tangram**.

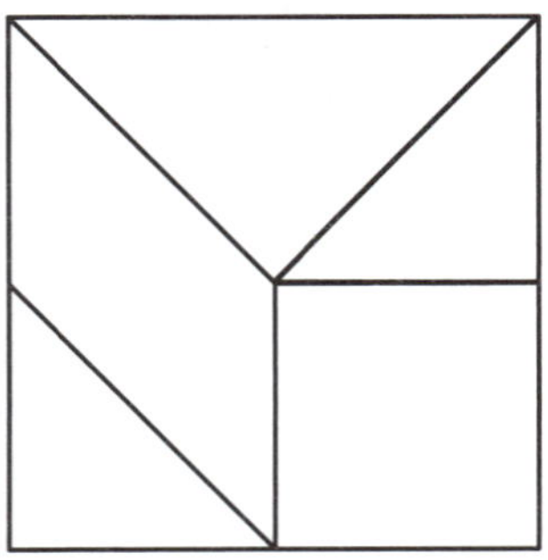

a Colour the square red.

b Colour the triangles green.

c One shape is not coloured. How many sides does it have? ☐

Colour this shape orange.

A tangram is an old Chinese puzzle.

 • *AUSTRALIAN SIGNPOST MATHS 1* • ISBN 978 1 4886 2178 9

26D The Cube

A face is a flat surface.

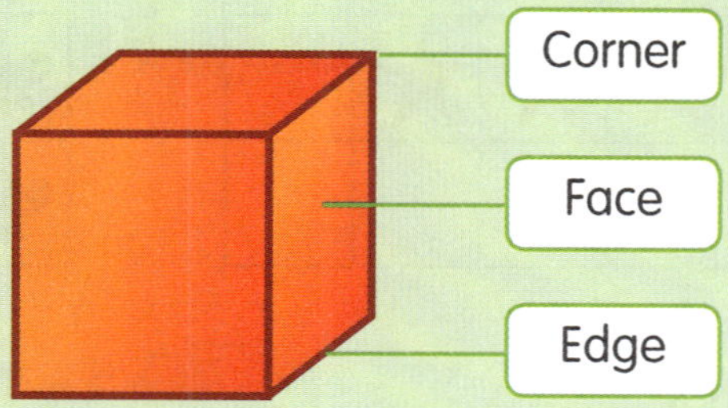

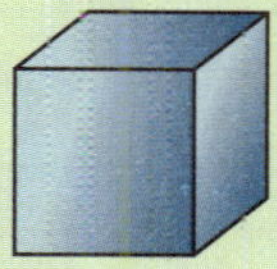

6 faces
8 corners
12 edges

1. Colour the cubes red.

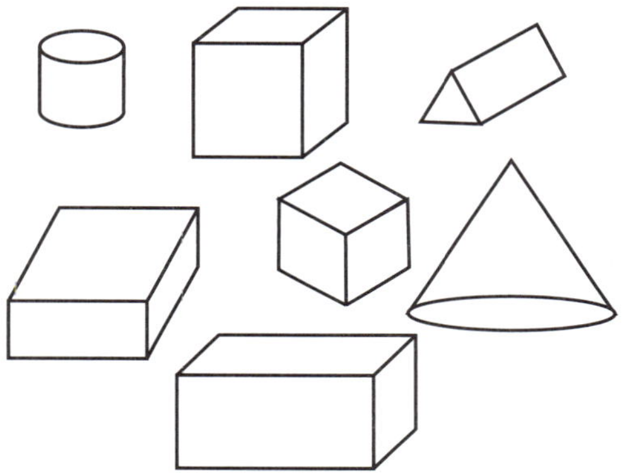

2. Circle the objects that can be stacked.

3. Describe a cube in words.

4. Build each model. How many cubes is it made of?

Do cubes stack well?

27A Subtraction by Counting On

1 Complete each line of pictures. Count on to find the answer.

a

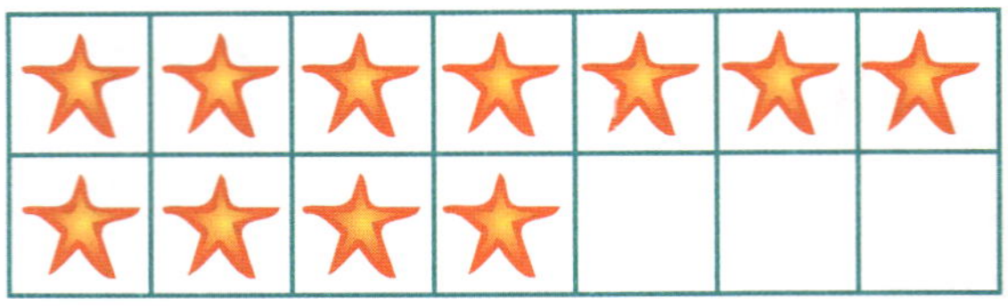

11 + ☐ = 14

14 – 11 = ☐

b

12 + ☐ = 14

14 – 12 = ☐

c

9 + ☐ = 12

12 – 9 = ☐

d

12 + ☐ = 16

16 – 12 = ☐

2 Complete the number sentences.

a

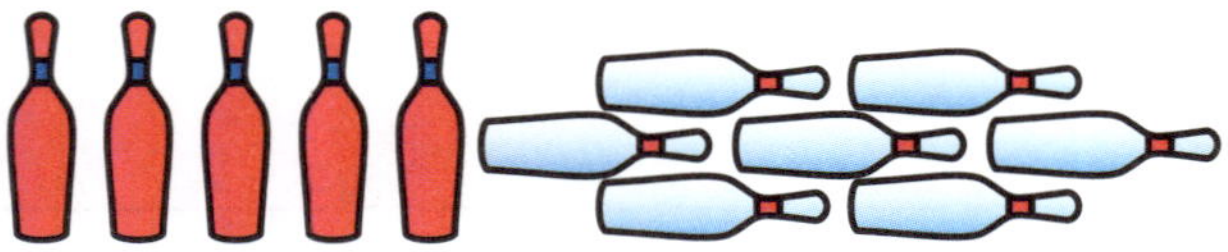

5 + ☐ = 12

12 – 5 = ☐

b

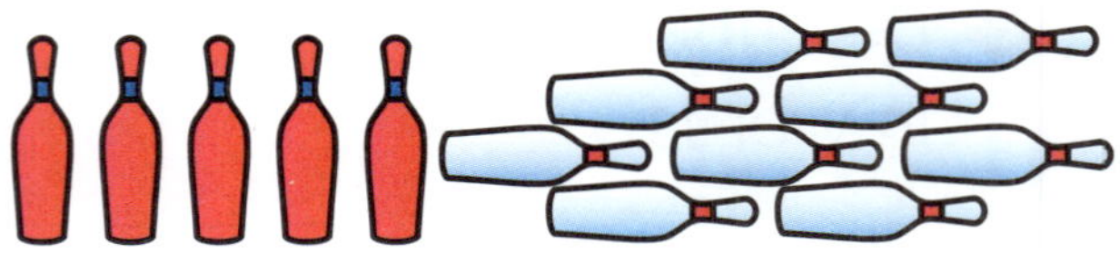

5 + ☐ = 14

14 – 5 = ☐

3 To find the answer, count on from the smaller number.

a 15 – 11 = ☐

b 17 – 14 = ☐

c 18 – 16 = ☐

d 19 – 15 = ☐

27B Addition

Sasha had some balls. She was given 3 more.
Now she has 10. How many did she start with?

- Colour 3 of the 10 balls orange.
- 7 balls are not orange, so she started with 7.

1. Kim had some ice creams. She bought 4 more.

   Now she has 6. She started with ☐.

2. Sam had some balls. He was given 3 more.

   Now he has 8. He started with ☐.

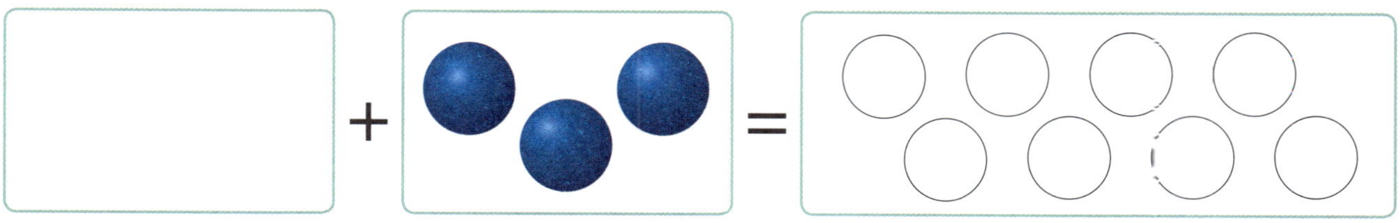

3. Melly had some fish. She bought 3 more.

   Now she has 9. She started with ☐.

 • *AUSTRALIAN SIGNPOST MATHS 1* • ISBN 978 1 4886 2178 9

27C One Half

❶ Circle the groups that are divided into halves.

a
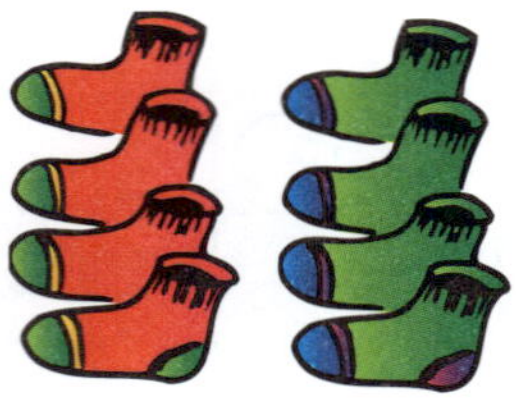

b
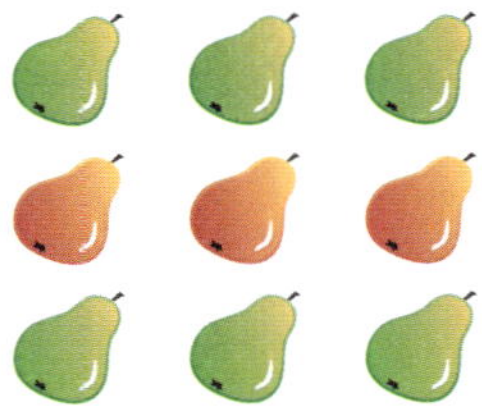

c
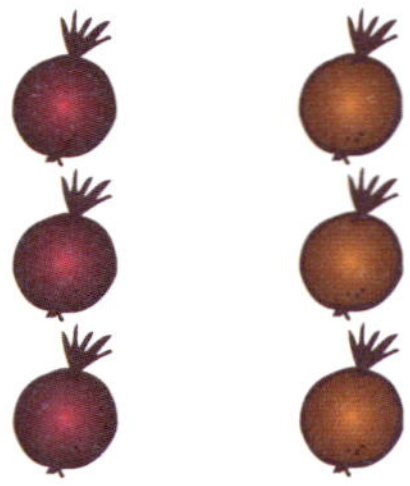

d

e

f

g

h

❷ Colour half of each group.

a

b
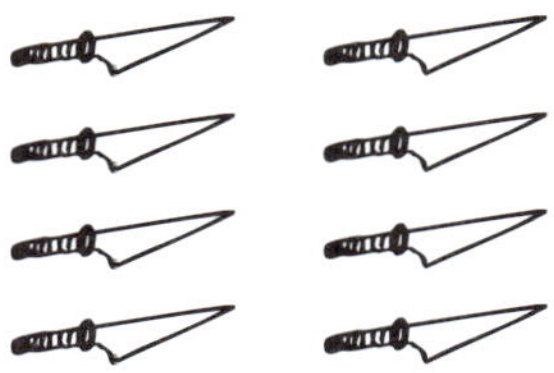

c
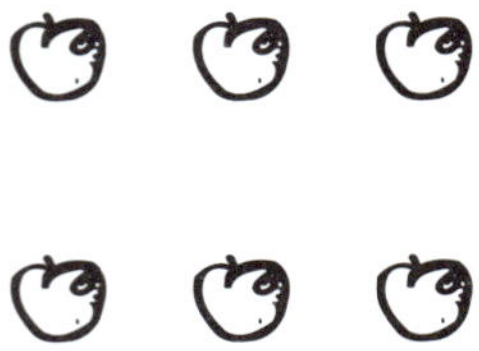

d

e

f

g

h

❸ Tick the correct box to show the best answer.

One half is $\frac{1}{2}$.

Group	Coloured Part		
	Less than Half	Half	More than Half
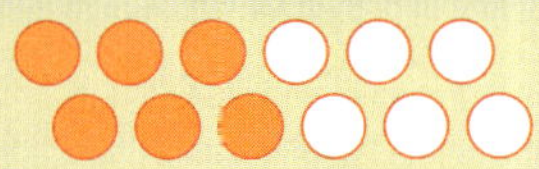			
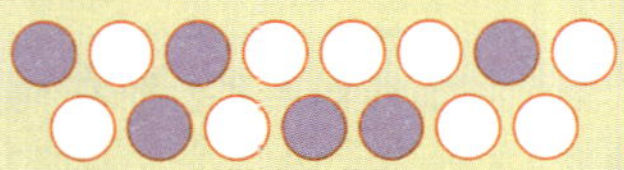			

27D Giving Directions

FUN SPOT

1 Use a counter to follow the instructions.

C			D
	E		
			B
	↑ Start		
		A	
G			F

1 Place your counter on the space marked Start.

2 Move the counter 3 spaces up. Colour that square red.

3 From the red square, move the counter 2 spaces right. Colour that square blue.

4 From the blue square, move the counter 5 spaces down. Colour that square green.

5 From the green square, move the counter 3 spaces left.

Where is the counter now?

2 Follow these directions from "Start" on the grid above. Where do you finish?

a 2 up, 2 right, 3 down, 1 left

b 1 up, 1 left, 3 down, 3 right

ACTIVITY

- Work with a partner.

  Take turns to give each other directions.

 • *AUSTRALIAN SIGNPOST MATHS 1* • ISBN 978 1 4886 2178 9

28A Place Value

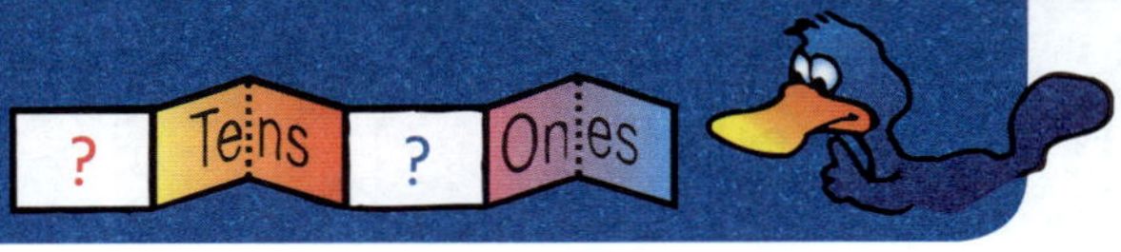

1 Write the number shown by the numeral expander.

a

b

c

d

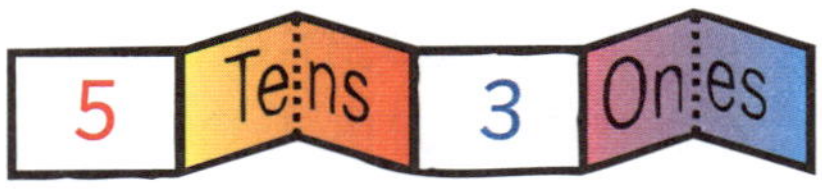

e

f

2 Expand the numbers.

a 97 = ☐ tens ☐ ones

b 59 = ☐ tens ☐ ones

c 60 = ☐ tens ☐ ones

30 = 3 tens 0 ones

3 Circle the larger number.

a 42 or 93 **b** 66 or 48 **c** 81 or 56

d 73 or 57 **e** 79 or 88 **f** 63 or 90

Work in pairs and use a calculator.

One person enters a number and hides the screen. The other person asks questions to try to guess the number.

Is it more than 50?

Is it less than 20?

 • *AUSTRALIAN SIGNPOST MATHS 1* • ISBN 978 1 4886 2178 9

28B Relating Addition and Subtraction

CONCEPT

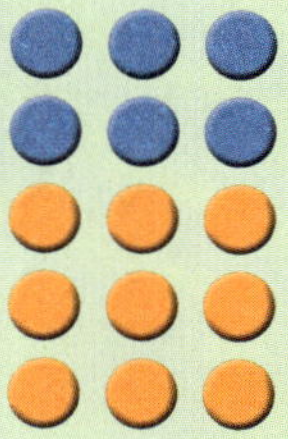

6 + 9 = 15

15 – 6 = 9   15 – 9 = 6

1 Complete the additions and subtractions.

a

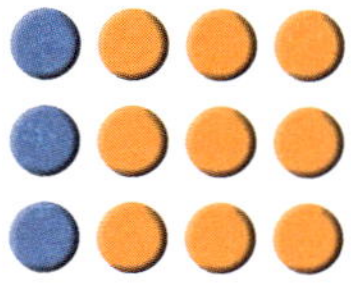

3 + 9 = ☐   12 – 3 = ☐   12 – 9 = ☐

b

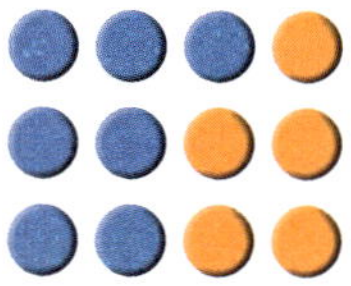

7 + 5 = ☐   12 – 7 = ☐   12 – 5 = ☐

c

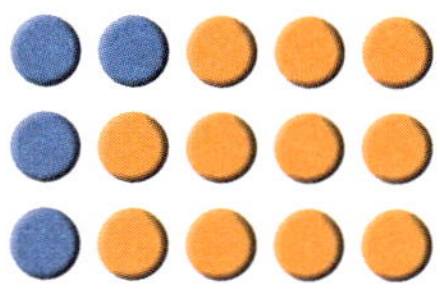

4 + 11 = ☐   15 – 4 = ☐   15 – 11 = ☐

2 Use the dots to write additions and subtractions.

a

☐ + ☐ = 10

10 – ☐ = ☐

10 – ☐ = ☐

b

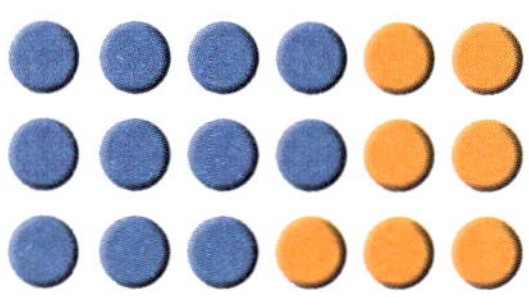

☐ + ☐ = 18

18 – ☐ = ☐

18 – ☐ = ☐

c

☐ + ☐ = ☐

☐ – ☐ = ☐

☐ – ☐ = ☐

28C Comparing Capacities

INVESTIGATION

1 Fill one container and pour the contents into the other.

Pan

Jug

The ______ holds more.

Container

Box

The ______ holds more.

2 Fill each container, pour it into a big measuring container and mark the level, then empty it. Compare the different marks.

Carton

Lunch Box

The ______ holds more.

Vase

Bottle

The ______ holds more.

3 How many cups fill each container?

Teapot

______ cups

The ______ holds more.

Lunch Box

______ cups

The ______ holds less.

28D Comparing Capacities

1 Compare the capacities of different containers.

Container Bottle

The ______ holds more.

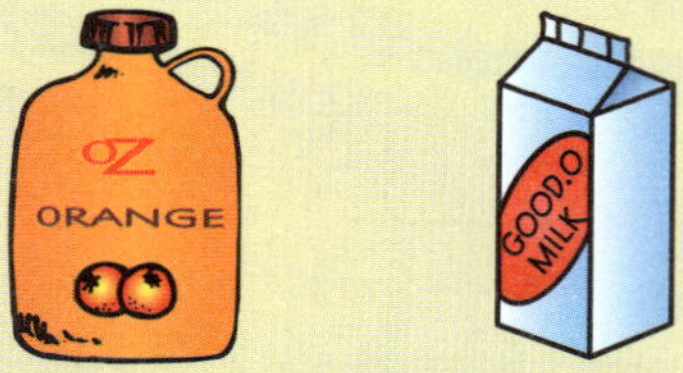

Bottle Carton

The ______ holds more.

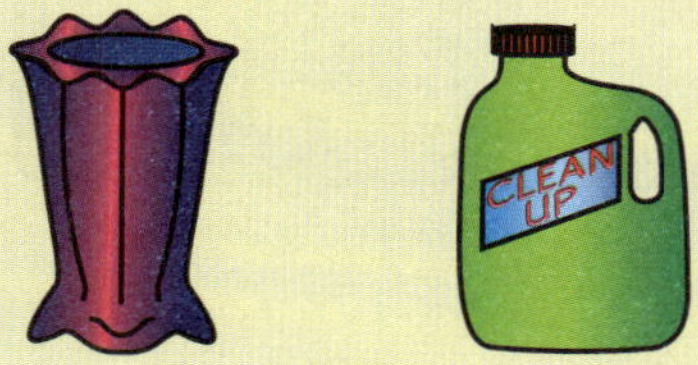

Vase Bottle

The ______ holds more.

2 Use a cup to fill each container.

Vase Glass Bottle

The ______ holds more.

The ______ holds less.

3 Use cups of water to compare three different containers.

Holds Most		Holds Least

1 Draw the missing dots on each domino to make the total. Complete each number sentence.

a  = 4

☐ + 2 = 4

b 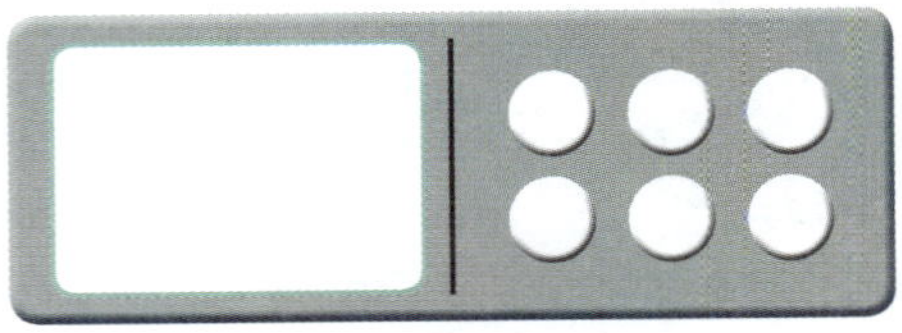 = 7

☐ + ☐ = 7

c 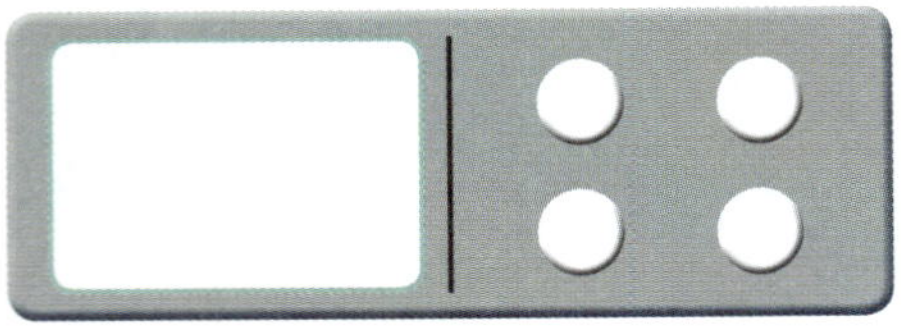 = 9

☐ + ☐ = 9

d  = 7

☐ + ☐ = 7

e  = 11

☐ + ☐ = 11

f 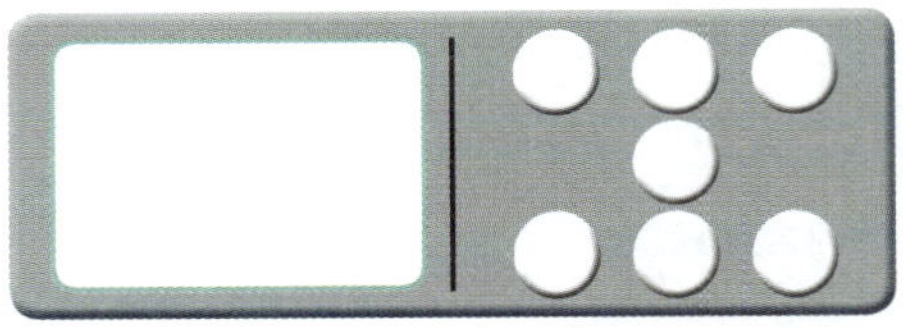 = 12

☐ + ☐ = 12

2 Sylvia had some stickers. She was given 4 more.
Now she has 12. How many did she start with? ☐

☐ ☐ ☐ ☐ ☐

 • *AUSTRALIAN SIGNPOST MATHS 1* • ISBN 978 1 4886 2178 9

29B How Many More?

Use this picture to answer the questions.

1 How many more do we need to make 10?

a footballs

7 + [ ] = 10

10 − [ ] = 7

b cones

[ ] + [ ] = [ ]

[ ] − [ ] = [ ]

c ropes

[ ] + [ ] = [ ]

[ ] − [ ] = [ ]

d children

[ ] + [ ] = [ ]

[ ] − [ ] = [ ]

e balloons

[ ] + [ ] = [ ]

[ ] − [ ] = [ ]

f bins

[ ] + [ ] = [ ]

[ ] − [ ] = [ ]

2 How many more footballs do we need to make 12? [ ]

29C Giving Directions

- Turn **clockwise** means turn from the top to the **right**.
- Turn **anticlockwise** means turn from the top to the **left**.

1 a Use this map and the words above to explain how to go from **B** to **F**.

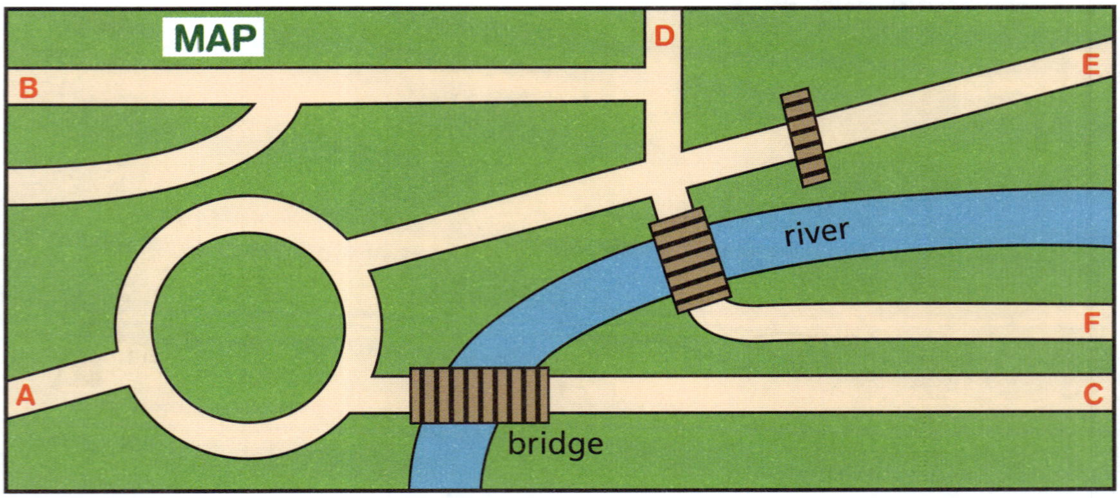

b Give directions to go from **A** to another letter on the map.
Work with a partner and take turns.

2

a If the girl goes forwards 3 squares, she reaches ☐.

b What forwards and backwards directions does she need to spell the word "CAKE"? Discuss.

3 Use directions to explain how to go to the principal's office.

 ISBN 978 1 4886 2178 9

29D Chance Language

1 For each picture, write **likely** or **unlikely**.

a I will stay home tonight.

b It will snow next week.

c An elephant will come to school.

d I will go to sleep tonight.

2 Draw something that will happen tomorrow.

3 Draw something that won't happen tomorrow.

 • *AUSTRALIAN SIGNPOST MATHS 1* • ISBN 978 1 4886 2178 9

30A Inverse Operations

Adding 5 is the opposite of taking away 5.

6 + 5 − 5 = 6

1. Rand had 10 balls. He found 6 more, then gave 6 away.
   How many does he have now?

   10 + 6 − 6 = ☐

2. Sarah had 12 stickers. She gave away 5, then was given 5 more.
   How many does she have now?

   ☐ − ☐ + ☐ = ☐

3. Perrin could see 12 dogs. Another 4 dogs came, then 4 left.
   How many can he see now?

   ☐ + ☐ − ☐ = ☐

4. **a** 12 + 3 − 3 = ☐ **b** 14 + 5 − 5 = ☐ **c** 13 + 5 − 5 = ☐

Create your own number stories and sentences where the addition and subtraction undo each other.

☐ ☐ ☐ ☐ ☐ ☐ ☐

 • *AUSTRALIAN SIGNPOST MATHS 1* • ISBN 978 1 4886 2178 9

30B Number Patterns

We can use the columns to add 10 or take away 10.

9, 19, 29, 39, ...
43, 33, 23, 13, ...

1	2	3	4	5	6	7	8	9	10
11	12	13	14	15	16	17	18	19	20
21	22	23	24	25	26	27	28	29	30
31	32	33	34	35	36	37	38	39	40
41	42	43	44	45	46	47	48	49	50

1 Write the next two numbers in each pattern.

a 4, 14, 24, ☐, ☐

b 7, 17, 27, ☐, ☐

c 45, 35, 25, ☐, ☐

d 14, 16, 18, ☐, ☐

e 50, 49, 48, ☐, ☐

f 3, 6, 9, ☐, ☐

2 Complete each number pattern. Describe the pattern.

a Can you see a pattern in the ones place?

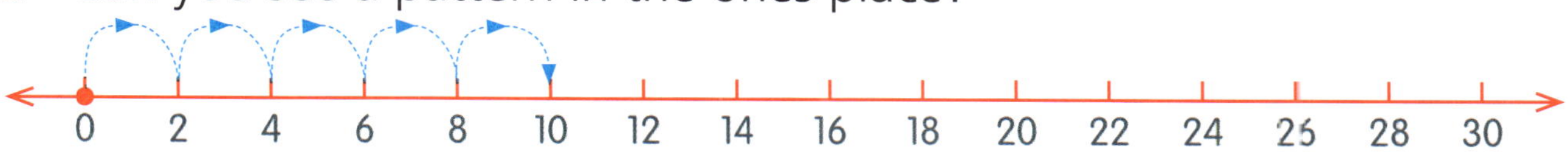

b Can you see a pattern in the tens place?

3 On a calculator, press these keys and write the answers you see.

7 6 [−] [−] 1 0 [=] [=] [=] [=] ...

☐ ☐ ☐ ☐ ☐ ☐

30C Duration Using Weeks

1 Discuss this timetable.

Week 1	Week 2	Week 3	Week 4	Week 5
Start of term	Sam's party	Start project	Zoo excursion	Test
Week 6	**Week 7**	**Week 8**	**Week 9**	**Week 10**
Class talks	Drama night	Project due	Parent–teacher interviews	End of term

2 It is now Week 5. Write what happened:

a four weeks ago

b two weeks ago

c last week

3 It is now Week 5. Write what will happen:

a next week

b three weeks from now

c four weeks from now

ACTIVITY

Make up your own term timetable. In your timetable, write the events that have happened and the events that will happen.

30D Duration Using Hours

1 Discuss this timetable.

7 o'clock	8 o'clock	9 o'clock	10 o'clock	11 o'clock
Wake up	Breakfast	School starts	Reading groups	Recess
12 o'clock	1 o'clock	2 o'clock	3 o'clock	4 o'clock
Maths	Lunch	Sport	School ends	Swimming lesson

2 It is now Recess. Write what happened:

a four hours ago

b two hours ago

c at 10 o'clock

3 It is now Recess. Write what will happen:

a two hours from now

b four hours from now

c at 4 o'clock

ACTIVITY

Make up a daily timetable of your own.

Talk about it with a friend.

 • *AUSTRALIAN SIGNPOST MATHS 1* • ISBN 978 1 4886 2178 9

31A The Calculator

[+] means **add**

[−] means **subtract**

[C] **clears** the calculator

1. Use a calculator to complete these.

a 6 [+] 4 [=] ☐ b 7 [+] 3 [=] ☐

c 8 [+] 2 [=] ☐ d 5 [+] 6 [=] ☐

e 3 [+] 9 [=] ☐ f 7 [+] 6 [=] ☐

2. Now try these.

a 9 [+] 0 [+] 8 [=] ☐ b 5 [+] 4 [+] 7 [=] ☐

c 6 [+] 4 [+] 5 [=] ☐ d 8 [+] 6 [+] 4 [=] ☐

e 7 [+] 10 [+] 3 [=] ☐ f 9 [+] 8 [+] 2 [=] ☐

- Make up calculator questions of your own.

☐ ☐ ☐ ☐ ☐

  • *AUSTRALIAN SIGNPOST MATHS 1* • ISBN 978 1 4886 2178 9

31B Calculators

1 On a calculator, press these keys and write the answers you see.

6 [+] [+] 10 [=] [=] [=] [=] …

2 Use a calculator. Discuss your answers.

a 0 [+] [+] 2 [=] [=] [=] [=] [=] [=]

Calculator shows:						
	2					

b 0 [+] [+] 3 [=] [=] [=] [=] [=] [=]

Calculator shows:						
	3					

3 Use constant addition (press [+] twice) to do these.

a 1 [+] [+] 10 [=] [=] [=] [=] [=] [=]

Calculator shows:						
	11					

b 5 [+] [+] 5 [=] [=] [=] [=] [=] [=]

Calculator shows:						
	10					

4 Use constant subtraction (press [–] twice) to do this.

92 [–] [–] 10 [=] [=] [=] [=] [=] [=]

Calculator shows:						
	82					

31C Mass

CONCEPT

- The lighter one goes up.

The bottle is heavier.

- The scales are balanced.

The apple and orange have the same mass.

1 Cross out the heavier object. Circle objects that are balanced.

a

b

c

d

e

f

g

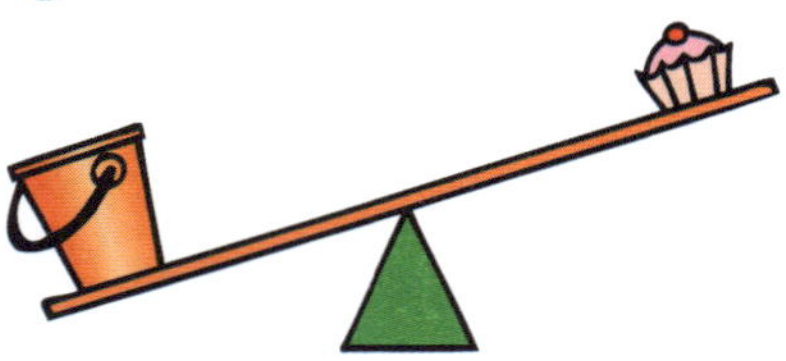

h

i

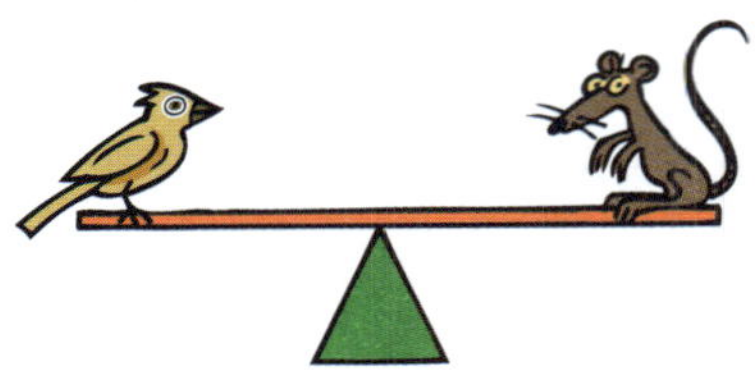

ACTIVITY

2 Use balance scales to compare these objects. Circle the heavier object.

 	 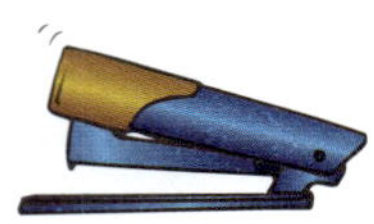

Which is heavier?

 • *AUSTRALIAN SIGNPOST MATHS 1* • ISBN 978 1 4886 2178 9

31D Dot Patterns

1 Complete the patterns.

a

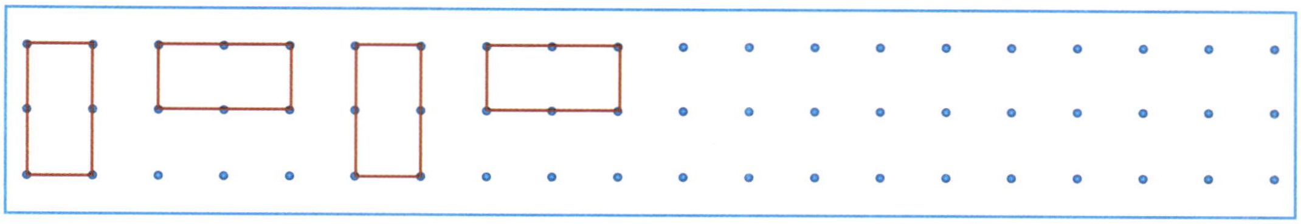

What shape has been used?

b

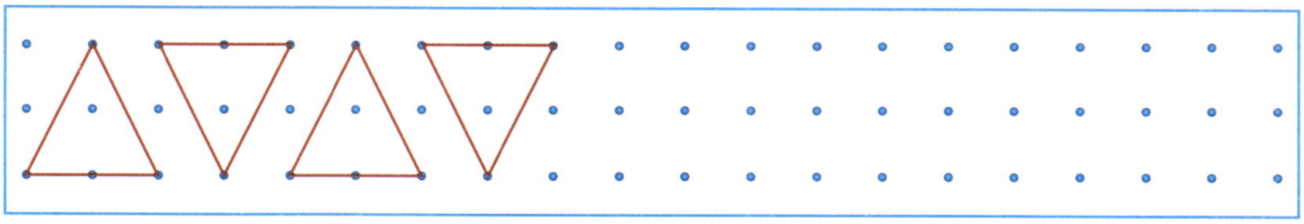

What shape has been used?

c

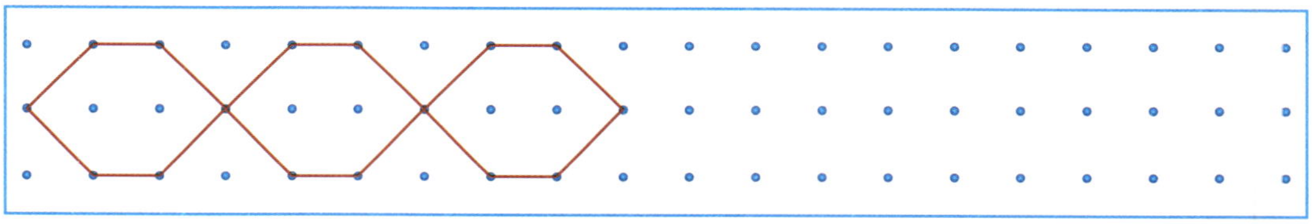

What shape has been used?

d

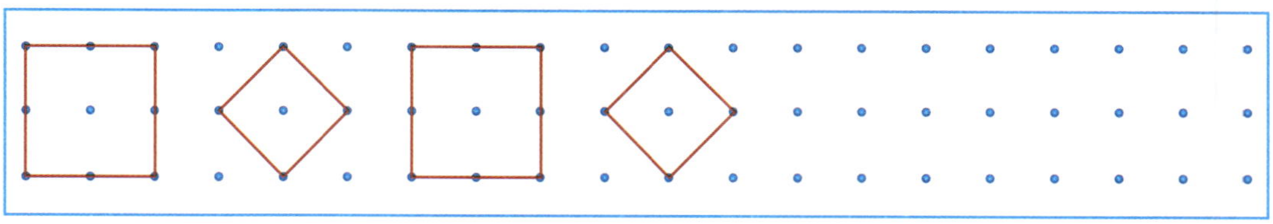

What shape has been used?

FUN SPOT

Draw your own shape pattern.

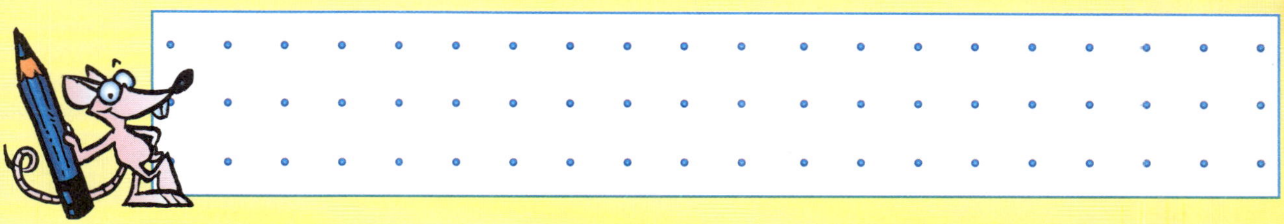

32A Ordinal Numbers

CONCEPT

1 Look at the picture above. Write the ordinal number word for the kangaroo that is:

a wearing shorts ☐

b the shortest ☐

c carrying a bottle ☐

d waving ☐

e wearing a hat ☐

f the tallest ☐

g wearing a shirt ☐

h coming last ☐

2 Circle these dates on the calendar.

a July 9th
b July 22nd
c July 3rd
d July 31st
e July 1st
f July 23rd

July

Sun	Mon	Tue	Wed	Thu	Fri	Sat
		1	2	3	4	5
6	7	8	9	10	11	12
13	14	15	16	17	18	19
20	21	22	23	24	25	26
27	28	29	30	31		

3 Colour these dates on the calendar.

a July 2nd
b July 7th
c July 19th
d July 30th
e July 25th
f July 5th

4 Write the ordinal number that comes next.

a 19th ☐

b 30th ☐

c 21st ☐

Appendix: Australian Money

1 Write the value of each coin.

2 Match each banknote to the correct label.

$20
twenty dollars

$50
fifty dollars

$100
one hundred dollars

$10
ten dollars

$5
five dollars

3 Write the value of the banknotes in order from smallest to largest.

1 How many?

a

b

c Fill in the missing numbers.

1			4				8		
		13		15		17			

2 Circle groups of 10.

3 Circle the:

a 6th

b 8th

4 Write the number.

a

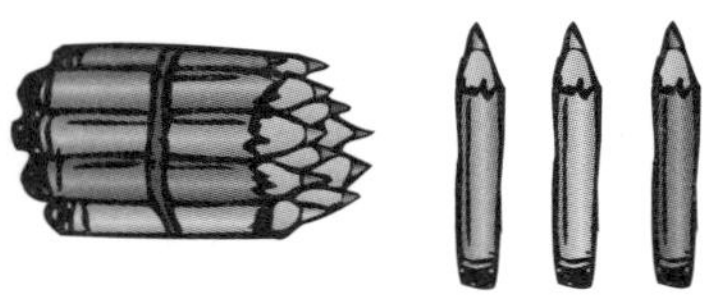

b

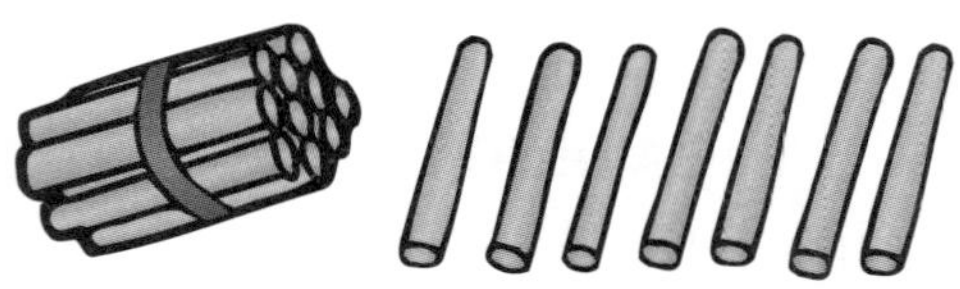

5 Complete:

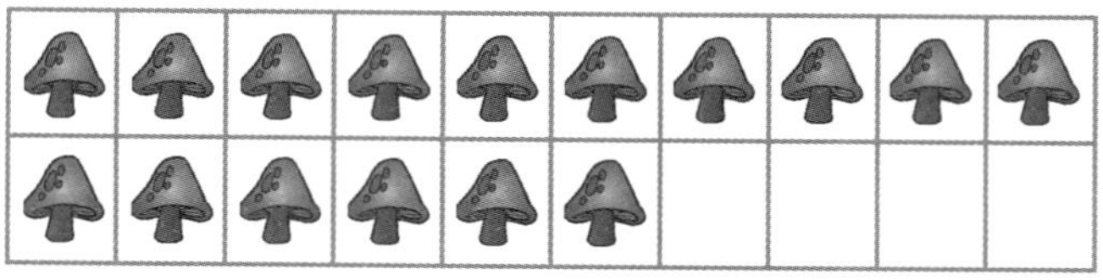

10 and 6 makes ☐

6 Complete:

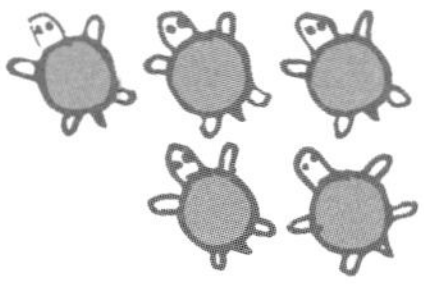

plus

equals ☐,

so ☐ + ☐ = ☐

7 Complete:

☐ take away ☐

is equal to ☐

8 **a** Colour one half of each.

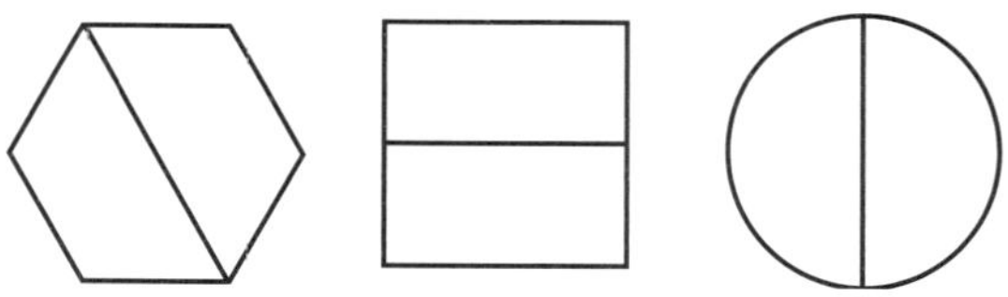

b Draw a line to cut the shape in half.

9 Write the time.

a

☐ o'clock

b

☐ o'clock

Show the time for:

c

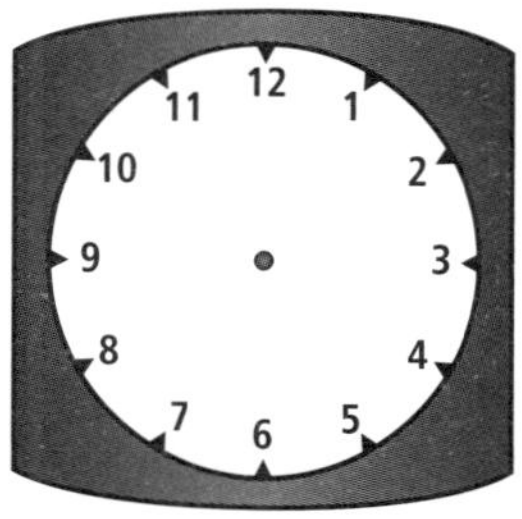

2 o'clock

d

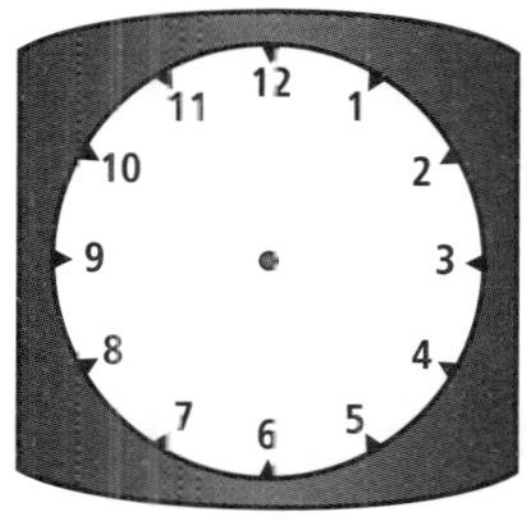

7 o'clock

 • *AUSTRALIAN SIGNPOST MATHS 1* • ISBN 978 1 4886 2178 9

10 Draw lines to match.

square | rectangle | triangle | circle | oval

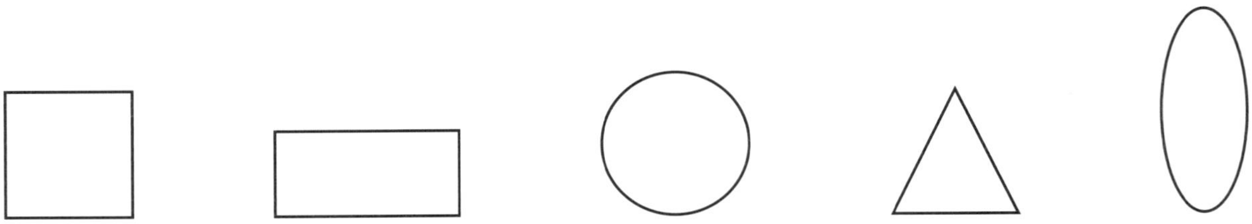

11 Draw lines to match the shapes.

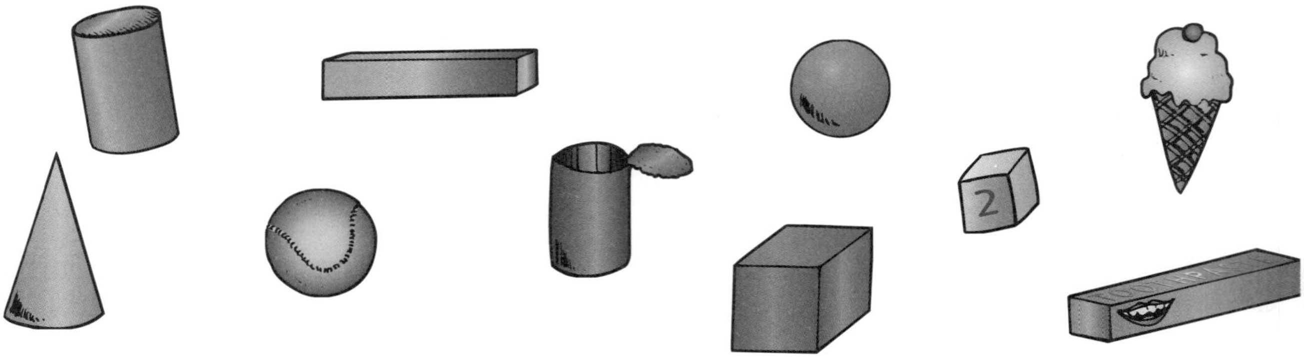

12

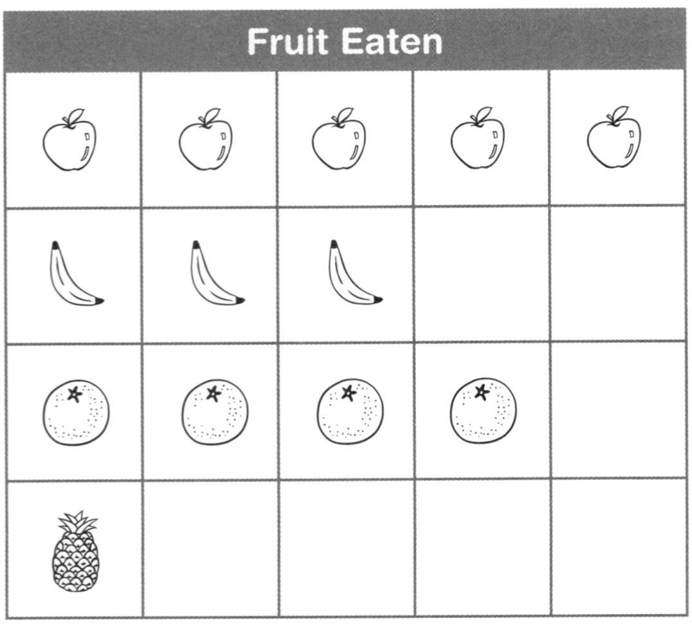

a Which fruit was eaten the most? ☐

b Which fruit was eaten the least? ☐

c How many 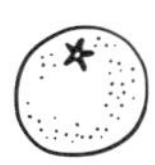 were eaten? ☐

d How many pieces were eaten altogether? ☐

Progress Test 2

1 Write the numbers before and after.

a ☐, 7, ☐ **b** ☐, 11, ☐

c ☐, 15, ☐ **d** ☐, 19, ☐

2

a 3 + 7 = ☐ **b** 4 + 5 = ☐ **c** 6 + 2 = ☐

d 8 + 1 = ☐ **e** 2 + 2 = ☐ **f** 4 + 4 = ☐

3

a 10 take away 2 = ☐

b 10 – 6 = ☐

4

a 14 – 6 = ☐ **b** 14 – 8 = ☐

5

☐ – ☐ = ☐

There are ☐ more than .

The difference is ☐

6 Count the groups of ten.

a 

☐ tens

b 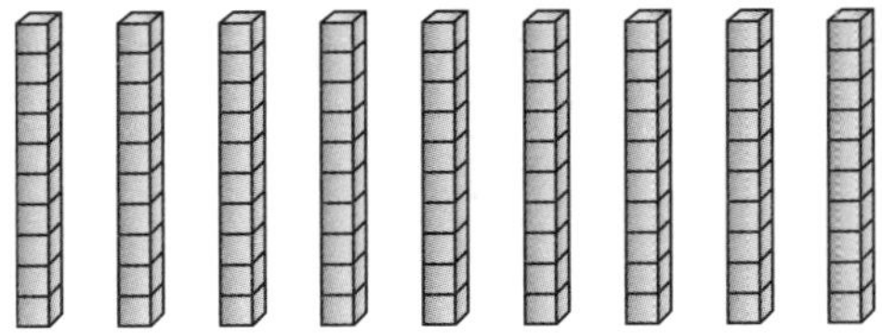

☐ tens

7 Complete:

a 10, 20, 30, ☐, ☐

b 50, 60, 70, ☐, ☐

8 **a** How many of each coin?

5c ☐ 10c ☐ 20c ☐

b Colour the gold coins yellow.

9 Complete the number pattern. Write how many to add each time.

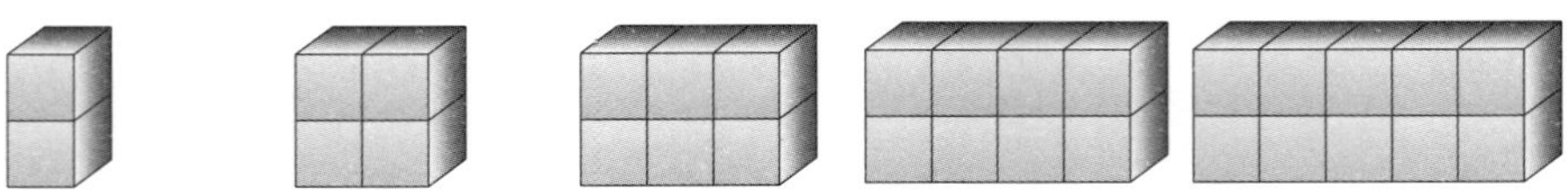

☐, ☐, ☐, ☐, ☐

Add ☐ is the rule.

 • *AUSTRALIAN SIGNPOST MATHS 1* • ISBN 978 1 4886 2178 9

10 **a** Circle the shortest pencil.

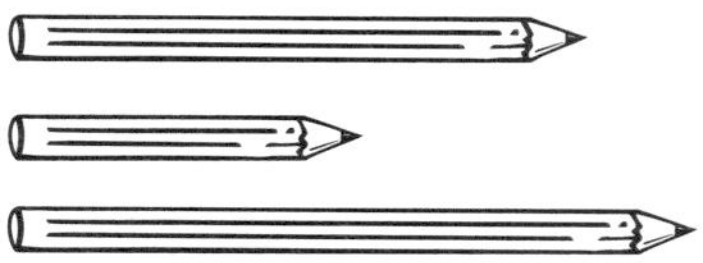

b Circle the longest snake.

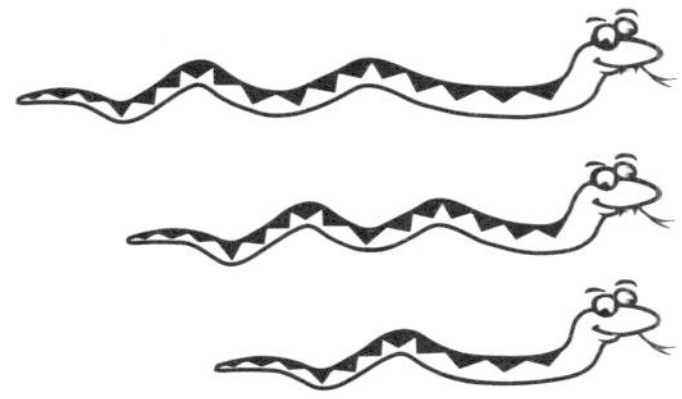

11 **a** Circle the heavier object.

b Circle the object with the greatest mass.

12 Write the time shown.

a

☐ o'clock

b

☐ o'clock

Show the time for:

c

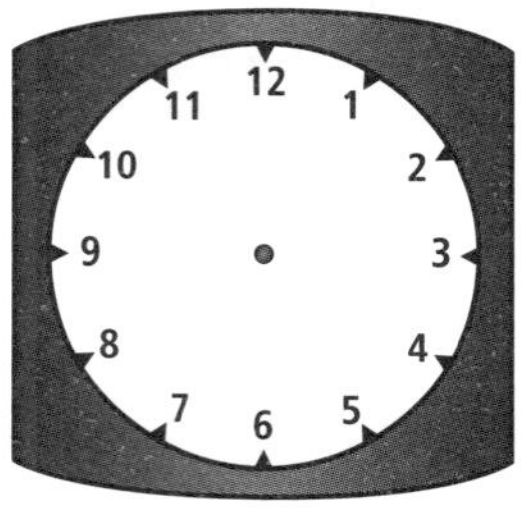

3 o'clock

d

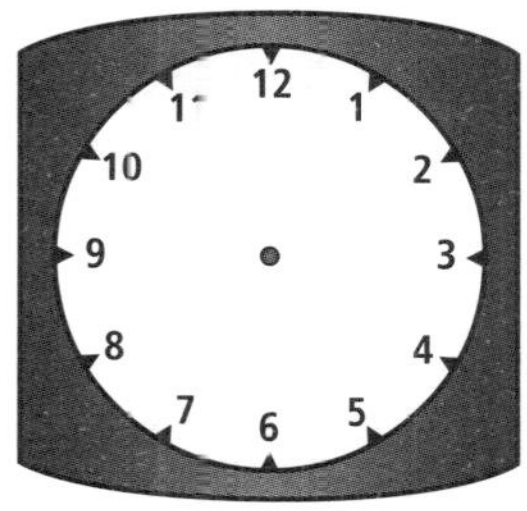

6 o'clock

13 How many of each shape can you see?

a triangles ☐

b squares ☐

c rectangles ☐

d ovals ☐

 • *AUSTRALIAN SIGNPOST MATHS 1* • ISBN 978 1 4886 2178 9

14 Colour the hexagon red and the oval blue.

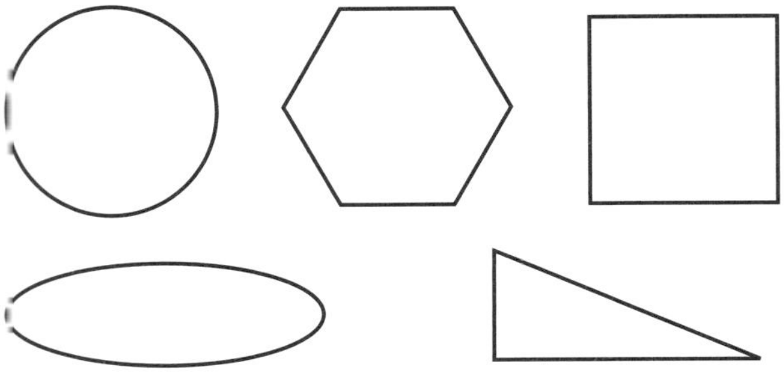

15 Colour items and shapes to match.

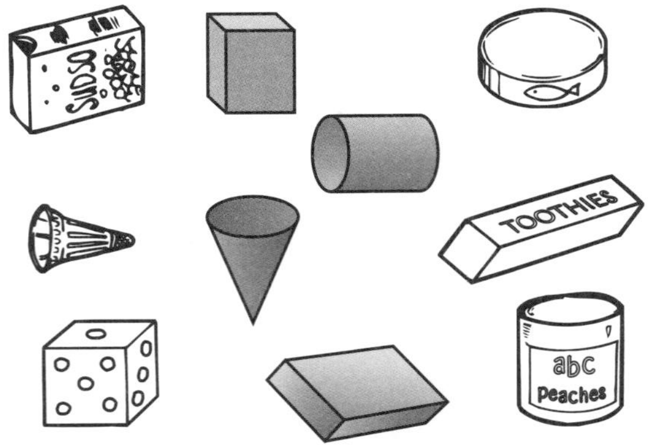

16 Write the missing word to show left and right.

a The ☐ is on the right.

b The clown is on the ☐

17 Look at the picture and circle the correct word in each sentence.

a The caterpillar is crawling along / beneath the stick.

b The mouse is above / below the hawk.

c The house is between / inside the hills.

d The hills are far from / close to the tree.

e The car is in front of / behind the house.

 • *AUSTRALIAN SIGNPOST MATHS 1* • ISBN 978 1 4886 2178 9

1. Write the numbers 12 to 19 in order.

[ ] [ ] [ ] [ ] [ ] [ ] [ ] [ ] 20

2. Count on to find how many altogether.

a 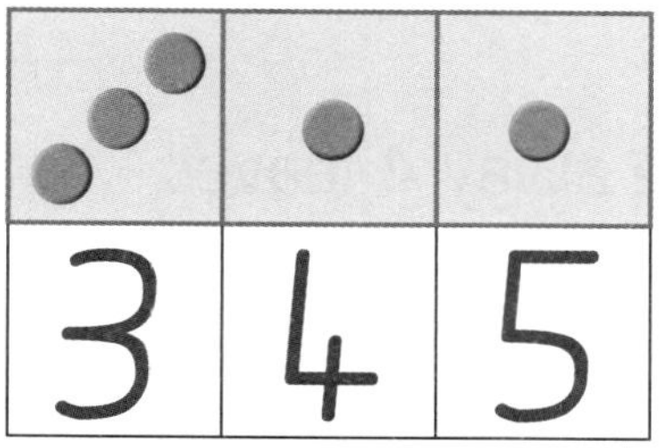

3 and 2 more = [ ]

b 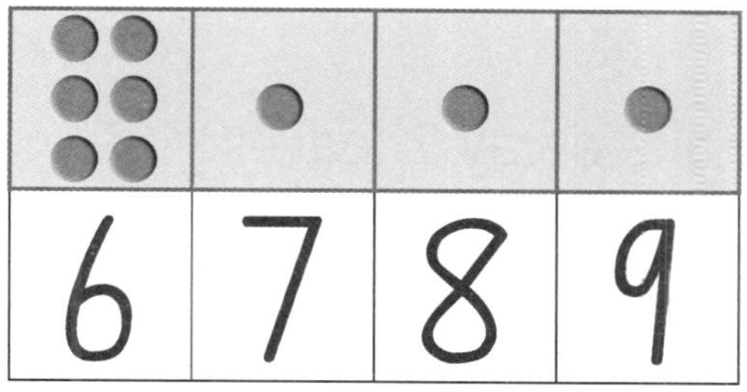

6 and 3 more = [ ]

3. Complete the number sentences.

a  and 

8 + 6 = [ ]

b  and 

[ ] + [ ] = [ ]

c  and 

[ ] + [ ] = [ ]

4. Colour the circles to show two different ways to make 9.

[ ] + [ ] = 9

[ ] + [ ] = 9

5 Count back to complete each number sentence.

a

8 take away 2 leaves ☐

b

11 take away 4 leaves ☐

6 **a**

14 – 9 = ☐

b

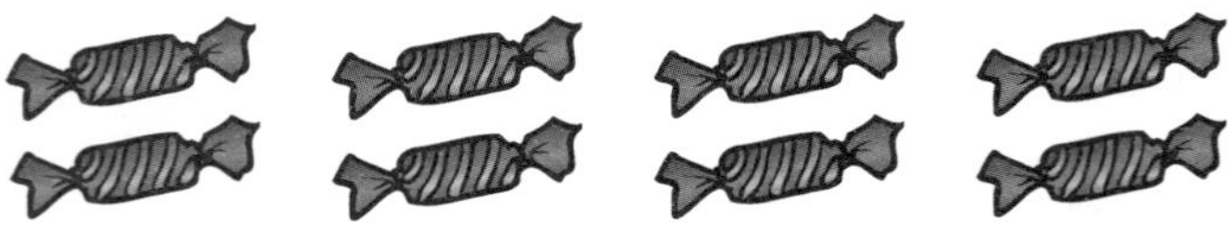

How many more make 12? ☐

c 8 + ☐ = 12

d 12 – 8 = ☐

e

19 – 8 = ☐

7

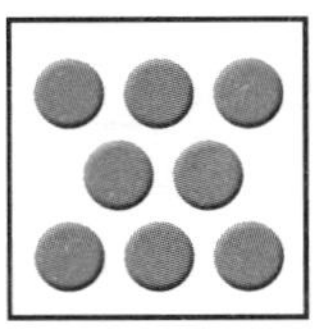 plus 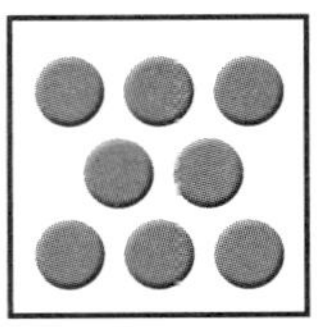 plus 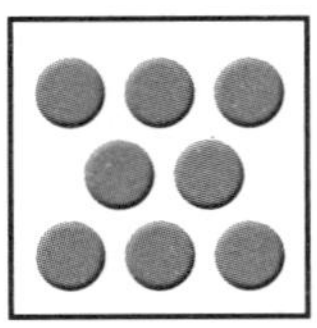

a ☐ + ☐ + ☐ = ☐

b ☐ groups of ☐ = ☐

8 Write the numbers before and after.

a

Before		After
	28	
	42	
	84	

b

Before		After
	53	
	76	
	91	

9 Write the numbers shown on each number line.

a

b
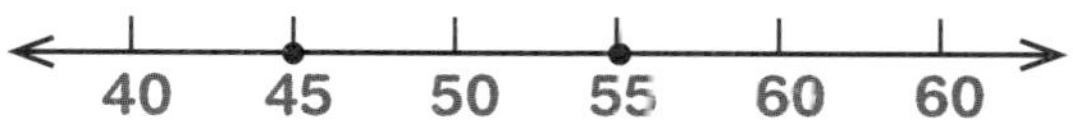

Show the numbers on each number line.

c 74, 75, 77, 79

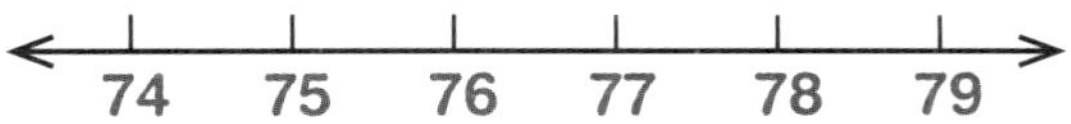

d 88, 90, 91, 92

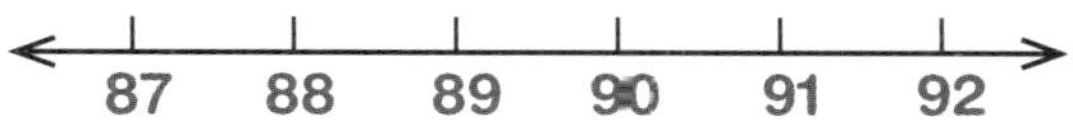

10 Complete:

a
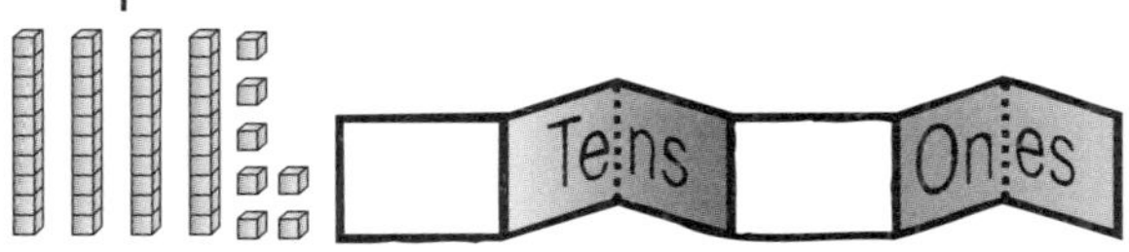

b
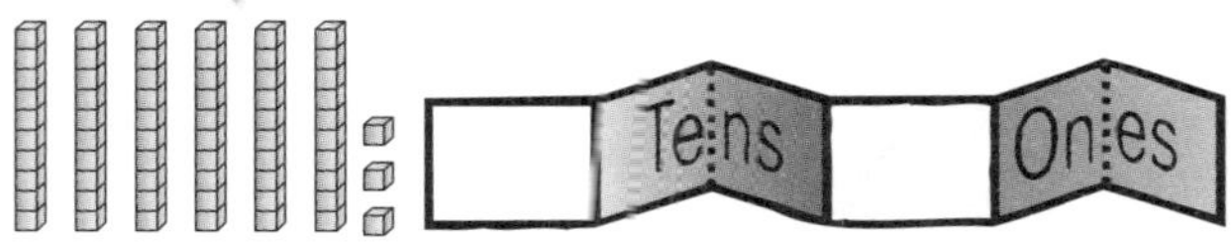

Write the number.

c

d

 • *AUSTRALIAN SIGNPOST MATHS 1* • ISBN 978 1 4886 2178 9

11 Write the numeral for each number.

a thirty-seven ☐

b sixty-eight ☐

c forty-three ☐

d Draw lines to match.

80	thirty-six
49	eighty
36	forty-nine

12 Complete:

a

5 | 10 | 15 | ☐ | ☐

b

22 | 24 | 26 | ☐ | ☐

13 Circle the object that will hold the least.

14 Circle half of each group.

a

b

15 Breakfast is at ☐ o'clock.

16

a

half past ☐

b

half past ☐

c

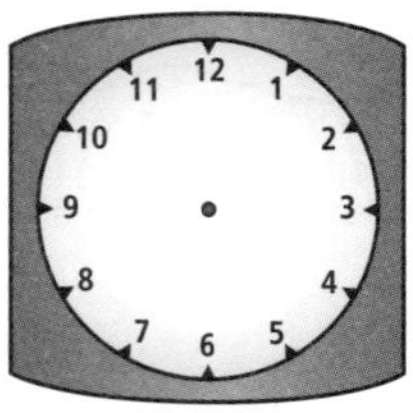

half past 1

d

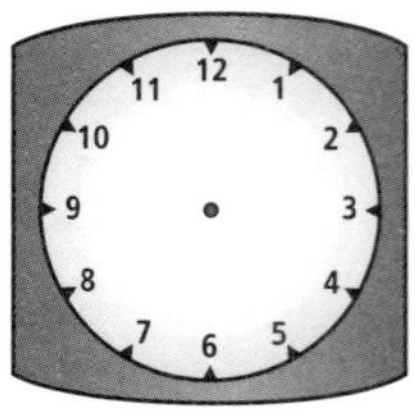

half past 10

 ISBN 978 1 4886 2178 9

17 Write the time.

a

half past ☐

b

half past ☐

c

half past ☐

Write the digital time for:

d half past 3 ☐

e half past 5 ☐

18 Name the two shapes of the faces.

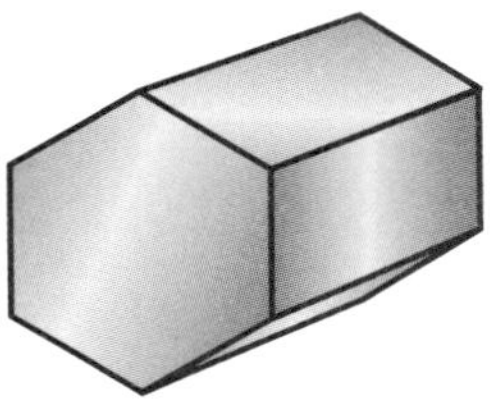

☐

☐

19 Complete the graph.

Shapes in the Picture

○								
□								
△								
▭								

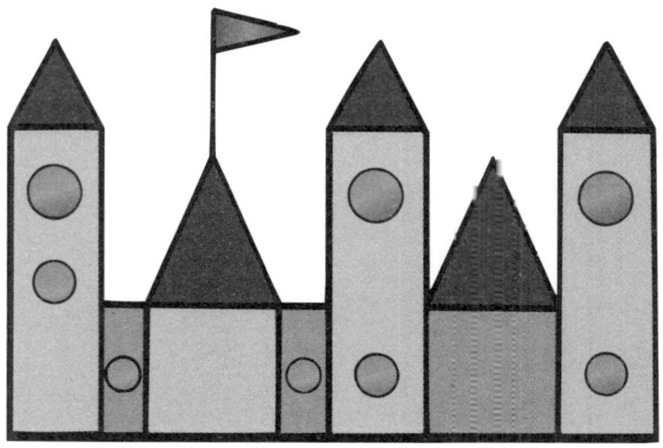

20 Write **will happen, won't happen** or **might happen** for each sentence.

a The sun will rise tomorrow. ☐

b A unicorn will be in my class. ☐

Progress Test 4

1 Draw dots in the ten frames to answer the questions.

a

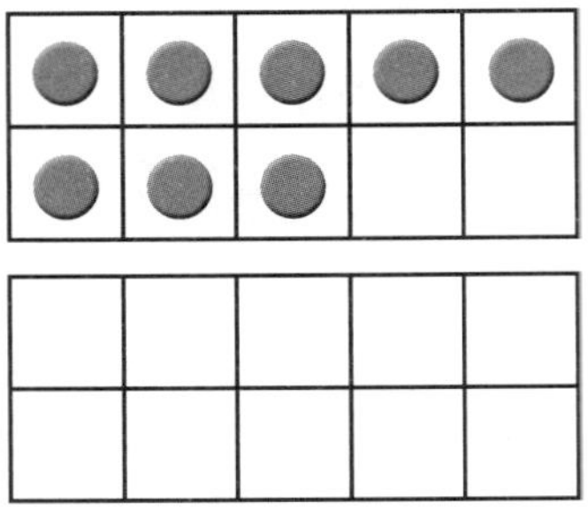

8 + 8 = ☐

b

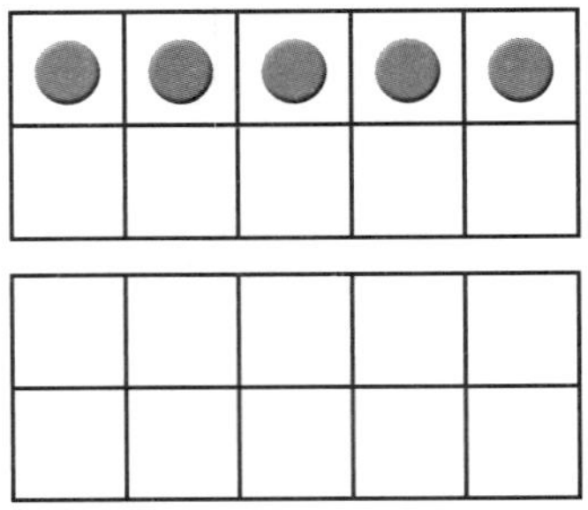

5 + 9 = ☐

c

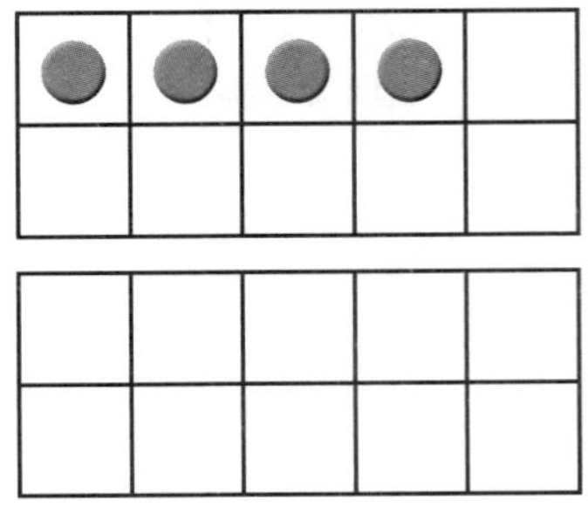

4 + 7 = ☐

2 Write the number.

a 9 Tens 2 Ones ☐

b 4 Tens 6 Ones ☐

c 8 Tens 3 Ones ☐

3 Complete:

a 78 = ☐ tens ☐ ones

b 80 = ☐ tens ☐ ones

4 What number is shown on each abacus?

a

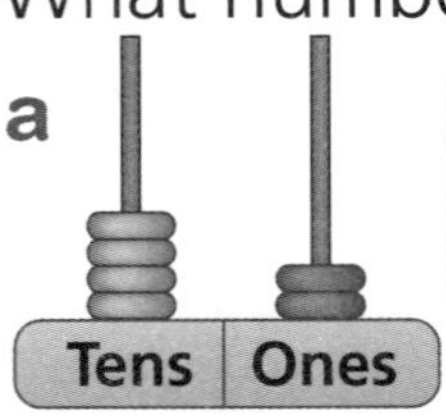

☐

b

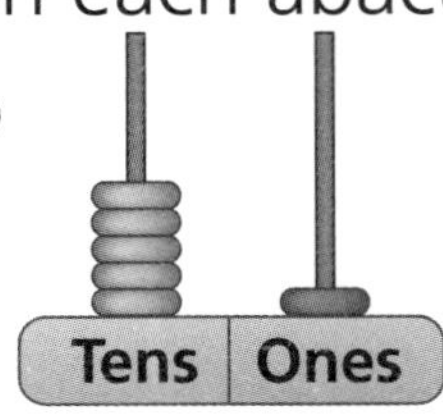

☐

c

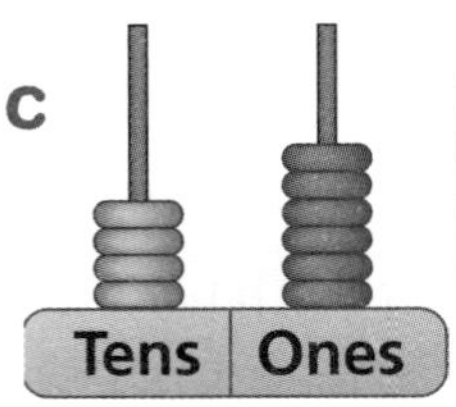

☐

5 **a** 16 + 3 = ☐ **b** 5 + 7 = ☐ **c** 8 + 6 = ☐

6 Kym had some ice creams. She bought 4 more. Now she has 7.

She started with ☐.

7 Complete the number sentences.

a

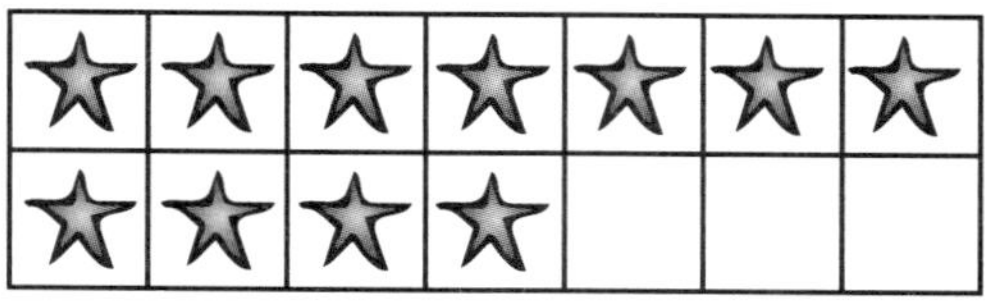

11 + ☐ = 14

14 – 11 = ☐

b

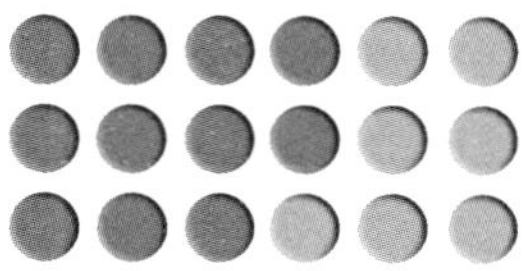

11 + ☐ = 18

☐ – ☐ = ☐

☐ – ☐ = ☐

8 Use the number line to complete the number sentences.

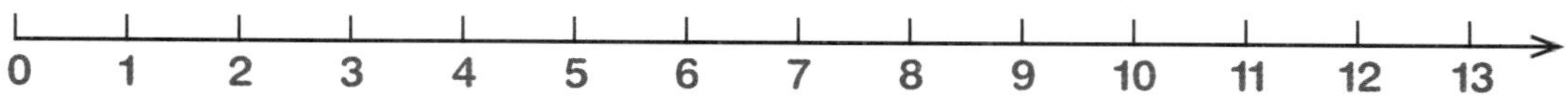

a 7 – 4 = ☐ **b** 13 – 5 = ☐ **c** 11 – 7 = ☐

9 Circle half of each group.

a

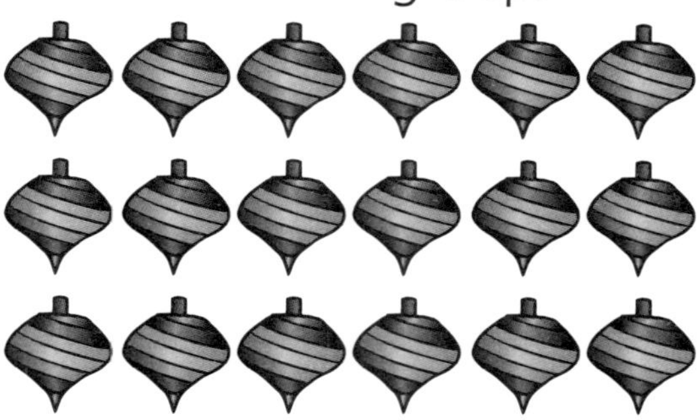

b

10 Complete the number patterns.

a 17, 18, 19, ___, ___, ___, ___, ___

b 12, 14, 16, ___, ___, ___, ___, ___

c 5, 8, 11, ___, ___, ___, ___, ___

11 Name two objects you could use to measure the length of your desk.

12

Cup

Vase

The ___ holds the most.

13 **a** How many months make one year?

b How many months make half of a year?

14 Write the time.

a

half past ___

b

half past ___

Write the digital time for:

c half past 12

d half past 6

 • *AUSTRALIAN SIGNPOST MATHS 1* • ISBN 978 1 4886 2178 9

15 Write or draw the time.

a

half past 11

b

half past 3

c

half past ☐

16
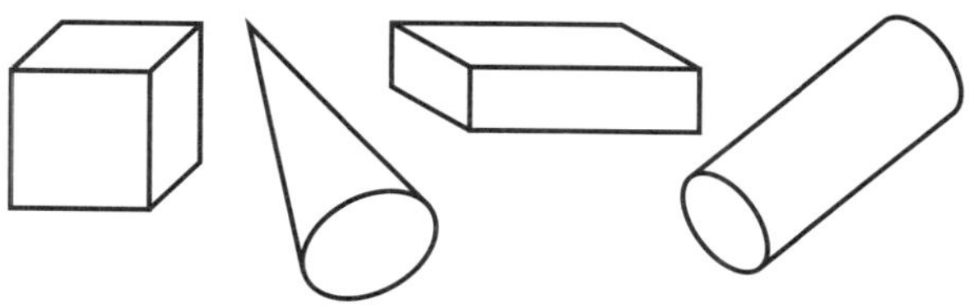

a Colour the cube red.
b Circle the objects that can be stacked.

17 Colour each rhombus.

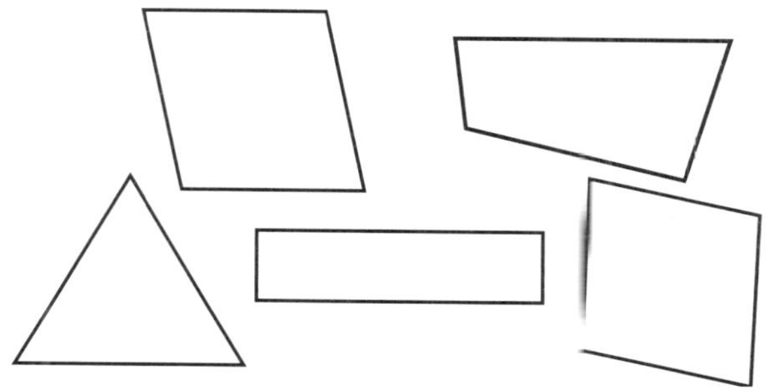

18 Complete:

Shape	Number of Corners	Number of Sides
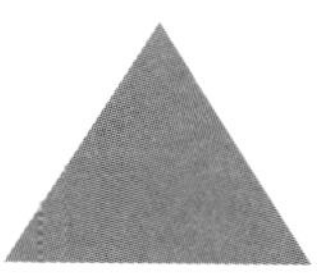		

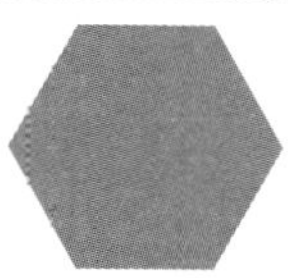		

19 Flowers Picked

Yellow	
Green	
Purple	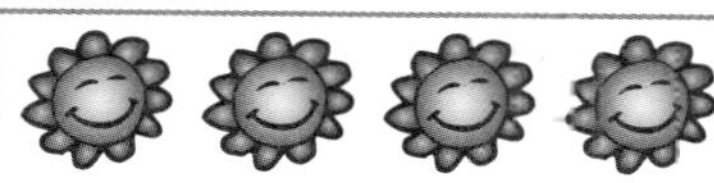

a How many yellow flowers were picked? ☐

b How many flowers were picked altogether? ☐

c Which coloured flowers were picked the most? ☐

 ISBN 978 1 4886 2178 9